ROUTLEDGE LIBRARY EDITIONS: POLLUTION, CLIMATE AND CHANGE

Volume 17

WHEELS OF PROGRESS?

WHEELS OF PROGRESS?

Motor transport, pollution and the environment.

Edited by
J. ROSE

Routledge
Taylor & Francis Group

LONDON AND NEW YORK

First published in 1973 by Gordon and Breach Science Publishers

This edition first published in 2020
by Routledge
2 Park Square, Milton Park, Abingdon, Oxon OX14 4RN

and by Routledge
52 Vanderbilt Avenue, New York, NY 10017

Routledge is an imprint of the Taylor & Francis Group, an informa business

© 1973 Gordon and Breach, Science Publishers Ltd

British Library Cataloguing in Publication Data
A catalogue record for this book is available from the British Library

ISBN: 978-0-367-34494-8 (Set)
ISBN: 978-0-429-34741-2 (Set) (ebk)
ISBN: 978-0-367-36489-2 (Volume 17) (hbk)
ISBN: 978-0-367-36493-9 (Volume 17) (pbk)
ISBN: 978-0-429-34638-5 (Volume 17) (ebk)

Publisher's Note
The publisher has gone to great lengths to ensure the quality of this reprint but points out that some imperfections in the original copies may be apparent.

Disclaimer
The publisher has made every effort to trace copyright holders and would welcome correspondence from those they have been unable to trace.

WHEELS OF PROGRESS?

Motor Transport, Pollution and the Environment

Edited by

J. ROSE

GORDON AND BREACH SCIENCE PUBLISHERS

London　　　　　New York　　　　Paris

The articles in this book were first published in *The International Journal of Environmental Studies*, Volume 3, Numbers 3 and 4, and Volume 4, Numbers 1 and 2.

CONTENTS

PREFACE

THE MOTOR VEHICLE is now occupying a commanding position in our society, particularly in the urban environment. It has become, in view of many, a master of our lives, though parading in the guise of our servant. The internal combustion engine has brought about a revolution in our society and appreciably altered the environment. Apart from the vast difficulties engendered by pollution, the motor vehicle has created a host of social and environmental problems requiring for their solution vision and also analysis in depth. To name just a few, one has to consider the relation between public and private transport, the elimination of traffic from city centres, economic consequences arising out of changed transport conditions, urban motorways, the provision of fuel in the context of the world's energy situation, etc. It appears that at present *ad hoc* and haphazard solutions are being tried, while great cities are being overwhelmed by the ever-growing armies of cars and lorries.

No doubt, the motor vehicle has contributed to the prosperity of the twentieth century man; it has also injured the quality of his life. To solve these urgent and difficult problems of modern life, it is essential to carry out extensive studies in order to evaluate with sufficient precision the problems generated by the motor vehicle and find the necessary solutions, if at all possible. It is hoped that this book, based on papers published in *The International Journal of Environmental Studies*, will assist in this exercise.

The articles deal with the effect of the motor vehicle on the environment and life, and are divided into four main parts: (1) Transportation and the Environment; (2) Safety; (3) Legislation; (4) Social Issues; the contributors are American and British experts. The impact of the motor vehicle on the environment is considered in six papers by Gwilliam, Johnson-Marshall, Joyce, McIlroy, O'Flaherty and Proudlove. Particular stress is laid on the design of towns and vehicles, economic problems associated with these, the responsibility of planners, and the integration of transport planning and environmental planning at local, regional and national levels.

The second part comprises three papers (by Baerwald, Foster and Mackay) concerned with safety. The main points considered are accident information, remedial programmes and the emerging science of accident research; a paper by Foster on hybrid cushion cars completes this part.

Legislation forms the subject of part three, the authors being Bampton, Gouse and Weighell. The need is advocated to establish basic criteria on a global scale concerning the control of man's pollution of the environment. The field of the present state of legislation involving the motor vehicle is reviewed, particularly that in the U.S.A., and differences between attitudes of the U.S.A. and Europe are pointed out.

The final part of the book consists of three papers (Hutchins, Paaswell, and Wall); topics considered are the problems of personalized transport, the car as an important factor for recreation and leisure, and the poor in an auto-owning environment. Some further trends are indicated and solutions suggested.

The problems treated in this book are of great importance and urgency. Solutions are badly needed; to obtain the necessary answers to the pressing issues it is essential to evaluate the problems, coolly and rationally. Hopefully, this book will help in this vital task.

J. ROSE

EDITORIAL VIEWPOINTS

The motor vehicle is occupying an increasingly central position in everybody's life. It performs transportation and distribution tasks, the importance of which in daily life cannot be sufficiently appreciated. This is true not only on a national but also on an international scale.

Having mentioned the tremendous positive factors, I think it is very important to state as well that the motor vehicle also involves great dangers to society. It causes numerous accidents resulting in injuries and damages; it plays an active part in polluting the environment emitting gaseous pollutants, thereby injuring the well-being, comfort and possibly also the health of man. Furthermore, it produces noise, which in many cases, exceeds that which the public's well-being and health can tolerate.

Like other countries Sweden is quite aware of these negative factors. We are firmly determined to reduce them as far as possible. A great deal has been achieved already in order to improve road safety, the inner safety of motor vehicles, as well as the air pollution and traffic noise situation. Much more, however, remains to be done on a national basis as well as internationally, and will be done. Otherwise we would neglect our basic duties and seriously endanger the quality of our life.

I should like to stress one of the points much discussed at present in Sweden. The benefit of the regulations passed so far in Sweden and other European countries in order to reduce the emission of gaseous pollutants by motor vehicles will, no doubt, be annulled before long because of the steadily increasing number of motor vehicles. It is evident that powerful action to limit air pollution from motor vehicles is needed not only in the United States, where regulations are being made more and more stringent, but also in Europe, if the positive effect of the actions taken up to now is to be retained and strengthened in spite of the increasing motor vehicle density. The basic principle, however, should be that conflicting national regulations should be avoided as far as possible, and that international regulations with the broadest possible applicability should be worked out. Having this in mind a Swedish group of experts, appointed by the Government, has proposed that Swedish regulations in this field should be made stricter, and has thereby recommended a modification of regulations which can be internationally applied. As a modification by Sweden of EEC regulations now in force would not involve any improvement of air pollution, the group of experts suggests that the Swedish regulations be adapted to the federal U.S. regulations. It is the opinion of the group that by doing so it should be possible to obtain the advantages from the point of view of atmospheric pollution that the best technique in the field of emission control can give. It is proposed by the group of experts that, as a first step, Swedish regulations should be strengthened by adopting provisions which should mainly correspond to federal US regulations for 1973 models of light duty vehicles.

The aim of the Swedish Government, when considering the report by the group of experts, has been to avoid acting unilaterally and adopting stricter regulations in Sweden without having previously made all possible efforts in the international field to achieve a satisfactory strengthening of the EEC provisions. Thus Sweden recently proposed within the EEC that new EEC regulations should be prepared and that those regulations should, in principle, correspond to the federal US regulations for 1973 models of light duty vehicles. It is felt in Sweden that, so far, international efforts to limit air pollution from motor vehicles have been lacking sufficient environmental concern. Therefore, Sweden has also suggested within the EEC that international efforts should be broadened in order to give the environmental aspects full consideration and that environmental specialists should be given more influence in this field. It is to be hoped that it will be possible to reach a satisfactory international solution to the problem of air pollution from motor vehicles. In our view this problem is of vital importance.

BENGT NORLING
*Minister of Transport
and Communication, Sweden*

The cause of clean air permits no compromise. The threat of polluted air is the threat of human sickness and, in some cases, death. We must, consequently, assign high priority to our task of cleansing the atmosphere.

In most urban areas of the United States, the internal combustion engine is one of the greatest contributors to dirty air, and as such, is now under close scrutiny. We are now requiring that all new cars produced must meet new exacting standards for emission control of pollutants. And lengthy tests on all new cars have been conducted.

These emission control devices, by themselves, however, are not sufficient to do the job. Given the future continuing growth in the number of vehicles on the road, certain American cities are going to have to take further measures. The clean air standards set by the Federal government will necessitate their actually restricting the number of cars entering and operating on their city streets. We shall, in effect, be rationing the amount of clean air that can be allotted to the internal combustion engine.

This prospect must form the foundation of all future urban transportation planning. Our planning must insure that successful alternatives to the private auto will be available and operating. We must, furthermore, insure that these alternatives are available in the near future—before the restrictions on private vehicles go into effect. It is folly to assume our society will shift its allegiance from the private car to public transit overnight. This transition must be gradual. More importantly, it must begin now for our lead time—given the customs, preferences and institutions that must be changed—is very short.

JOHN VOLPE
U.S.A. Secretary of Transportation

PART ONE

Transportation and the Urban Environment

THE URBAN ENVIRONMENT AND THE MOTOR VEHICLE

PERCY JOHNSON-MARSHALL

Faculty of Social Science, University of Edinburgh, Edinburgh EH8 9YL, U.K.

This Paper makes a case for the comprehensive and integrated approach to the planning of the environment and the motor vehicle. It points out that cities have had transport problems for many hundreds of years, and that one of the basic causes is the inability of the fabric of cities to adapt in a sufficiently flexible way to meet the rapid advances in transport technology. If, however, requirements of vehicular transport are given precedence over other aspects of urban life then the resultant damage to many existing urban fabrics may be very serious.

It is necessary to consider the problem at all three planning levels, national, regional, and local, but in Britain this has not yet been attempted. It is also necessary for the public to have a clearer idea of needs and possibilities so that misunderstandings between planners and public may be dealt with.

A brief reference to urban evolution is made in order to see the problem in its evolutionary context. The impact on urban environments of the new technological systems of transport of canals, railways, and motor transport is referred to, as is the serious effect on American cities by the latter system.

The current trend of decentralization of individual urban components is mentioned, and a warning of its effect is given. A warning is also given of the influence of the large scale Transportation Studies introduced from America, and the danger of considering them in isolation. Too few public authorities have as yet given sufficient attention to a much more important British contribution, the Traffic in Towns Report by Professor Buchanan.

In conclusion, the importance of the integration of transport planning with environmental planning at national, regional, and local levels is stressed, as is the need to distribute equally land uses and traffic generators. Finally more integration of rail and road systems is necessary, together with more study as to how this could be done.

In every age which has boasted an urban society there have been pressures on the form of cities. Unless a society is entirely static, a process of change is taking place constantly in the urban fabric, endeavouring to adapt and adjust to new forms of social organization, new environmental standards, new arrangement of land and building uses, and new means of communication.

The basic difficulty is that the building of cities represents a commitment of capital and of physical equipment in the form of artefact of many kinds. Once commitment has taken place, the developers, whoever they be, are understandably reluctant to see the whole process repeated until the particular artefact has served its purpose for a reasonable economic period. With most urban artefacts there is usually a degree of flexibility, but often the pressures for change, owing to such phenomena as population increase, economic development, or technological innovation, demand more rapid reorganization than the normal urban fabric permits. This is, of course, not a new problem for town planners. In the time of Julius Caesar changes were made in the urban pattern of Rome[1] by widening streets such as the Via Lata in order to meet the increasing demands of transport, by building colonnades to provide protection from climatic conditions, and by issuing edicts to control the use of streets by commercial traffic in the daytime.

Nineteen hundred years later, cities and towns were faced with the impact of a new transport system in the railway, which created widespread havoc in many a fine urban environment.

If urban societies think that they have an unique problem of traffic in towns today, a glance back through urban history may console us that, in the words of Ecclesiastes "History merely repeats itself. Nothing is truly new; it has all been done or said before. What can you point to that is new? How do you know it didn't exist long ages ago? We don't remember what happened in those former times, and in the future generations no one will remember what we have done back here."[2]

Fundamentally the problem is one of approach, of values, and of human tolerance. It is necessary for urban societies to decide on the priorities of importance which they give to the various aspects and functions of the city, including those for transport.

3

To put it in an extreme form, it has been proposed that a motorway should be driven across Venice,[3] and that a motorway ring should be driven through the fabric of Edinburgh.[4] The High Street of Oxford and Trumpington Street in Cambridge are both obviously unsuitable for the present day requirements of motor traffic, and a powerful case could be made for "improving" them both in the interests of vehicular movement.

The majority of cities and towns have had the bulk of their urban fabric determined over many generations. Some parts of this fabric may by now be of substandard quality and be due for comprehensive renewal, and many of the older industrial cities of Britain may be cited as examples. On the other hand, where societies have built with nobility of intent, with conscious pride in civic quality, and have created urban environments of quality, the case for conservation would appear to be paramount. Europe abounds in historic cities with this kind of problem, whether it be Rome, Florence, Prague, Paris, Edinburgh or Bath.

In between the two extremes of replaceable substandard environments and irreplaceable areas of civilized quality, lies a large and somewhat indeterminate group of problem areas for the town planner. It is in these areas that too often the traditional narrowly based traffic engineering solutions are pushed through at heavy expense and at high environmental cost to the community.

If one considers how the problem should be approached, in a fundamental way, it is essential to go back to the grass roots of urban planning, and to ask what cities and towns are for, even whether they are still necessary at all. Assuming that the answer is yes, the problem has to be tackled at both extremes of the space scale, on the one hand by extending upwards from a study of human *per capita* activities and space needs, and on the other by extending downwards from a national environmental plan based on the optimum of use of resources of all kinds, including those for all kinds of transport. Below the national environmental plan would come regional plans, setting out the main regional land uses, settlement pattern, and integrated transport network. Within the regional strategies, city and town plans could begin to put forward sensible decisions regarding the urban environment, and within these plans various types of action plans for precincts or environmental areas could be developed and carried out on the basis of the *per capita* standards.

All this is familiar enough to any contemporary town planner, but the comprehensive nature of environmental planning needs to be stressed at all levels. Today in Britain there is still no national environmental plan that relates settlement distribution, industrial production and non-urban resources with communications, or integrates the rail, road and air transport systems into a broad environmental strategy. If it had existed it is highly unlikely that there would have been the £1 million farce of the third London Airport, the four consecutive attempts at planning the South-East of England, or the widespread closing of rail lines while the accidents build up on the primary road system. Similarly, there is still no recognizable pattern of regional plans, so that the highly organized local development plan system has no settlement strategy within which to work. How can a city decide by itself on its own population and industrial targets, and hence on a rational transport system?

If one looks at the problem through the other end of the telescope from the point of view of the individual, one sees "the public", as usual, wanting a whole group of contradictory facilities, comparable to a demand for lower Income Tax but increased defence expenditure. The ideal of daily travel convenience is that of the image of the average man falling out of bed into his car, and thence, at lightning speed to work, and out of his car to his office chair or workshop bench. This image, ludicrous as it may be in fact, is pushed hard by a wide range of powerful promoters, such as the car manufacturers and salesmen, the road makers, and a large number of ancillary promoters, even including cigarette advertisers. In fact, one of the main causes of frustration between planners and planned is the wish of the public, or of articulate sections of the public, for impossible, or for contradictory, objectives.

No more clearly are the contradictions observable than in the case of the motor vehicle, but it seems to be impossible to explain to the average man that it is well beyond the bounds of possibility to remodel an existing city on the basis of a total use of private vehicles, short of destroying a large part of the urban fabric. If one could obtain some measure of agreement on a few basic considerations, the resolution of some of the more intractable problems could begin in a more reasonable and realistic manner. Among these I would include the following:

—The traffic will always increase to fill the roads available.
—It is not possible for everyone capable of driving

to drive to all or even the majority of daily urban destinations in existing cities.

—It is not sensible to mix up cars and people at the same level in shopping streets and other urban conditions of high intensity of land use, i.e. town centres.

—It is not sensible to allow a large number of cars in town centres as they (1) can be a source of accidents and a health hazard in terms of atmospheric and noise pollution, (2) are a wasteful use of space both when in motion and at rest.

—It is not advisable to mix different kinds of transport on the same pavement (I use the word both with its British and American meanings).

Even before the motor car, and indeed as far back as Roman times, a degree of separation of vehicles and pedestrians was found to be necessary, in terms of roads and pavements or, in American terminology, pavements and sidewalks. In the nineteenth century it was at an early stage realized that the new rail system would require segregation (except in the case of many new settlements in the then developing countries). In the late nineteenth century the light railway, in terms of the tram, tried to break this excellent rule. Thanks to this stupid mistake the tram is almost extinct, although, like the train it could have great potential value for urban transport.

In the twentieth century the indiscriminate and largely uncontrolled use of the motor vehicle has forced drastic remedies of various kinds. One has been the first aid control system of closing certain streets to vehicular traffic, or of precinctual restrictions, so that parts of town are made largely into pedestrian islands, while another is that of vertical segregation, whereby a separate pedestrian system is created either above or below the traffic streets.

Perhaps a brief reference to urban evolution in relation to transport might at this stage be useful. Until the industrial revolution cities and towns were generally small in size and compact in form. Urban populations were small: the city was urban, often urbane, and was clearly visible as an entity in the seemingly limitless countryside. The larger cities, too, were usually on or near harbours or navigable rivers. Inter city transport was usually by boat or by horse, carriage or cart, the latter modes time-consuming and often extremely uncomfortable. It is difficult to imagine London being a week away from Edinburgh, when today one can travel to almost any large city in the world in less than half the time—

as for tomorrow, who knows? For the most part ex-urban travel was based on the city and town being the market centre for the surrounding rural area, so forming an early type of traditional regional-city complex

In so far as the urban areas were concerned, they were always congested by horse, carriage and cart traffic on market days and during festivals, but as they were small the problem was reasonably contained. In addition to the close proximity of the countryside, there were usually spacious squares, and, after the Renaissance, spacious new streets as well. Horse drawn traffic only became a large scale serious urban problem after the Industrial Revolution, when towns were expanding so rapidly, and then the canals and railways took off a good deal of the pressure.

During and after the Industrial Revolution, new methods of transport technology were an essential part of the whole process. Inter-city transport was first improved by a more scientific method of road making, so as to bring it approximately up to Roman or Inca standards. The first technological breakthrough was with the canals, whereby larger numbers of people and greatly increased goods could be transported inland, thus increasing considerably the potential location of settlements. The revolutionally breakthrough was, of course, with the railway. All the settlements of a country could now be linked by a rapid transport service capable of carrying large numbers of people and large quantities of goods.

The impact on the cities and towns was startling, and none was the same again. Indeed, trainless towns such as Stamford became anachronisms until the arrival of the motor vehicle. The railway brought a continuous and segrated network of highly controlled linear strips into the existing urban fabrics, and included the urban components of stations (as new urban thresholds), tunnels, bridges, and large unsightly shunting yards. With its noise and smoke and lack of respect for traditional environmental quality, it was at first regarded, especially by intellectuals like Ruskin, as a de-civilizing agency, and at best tolerated as a necessity.

It was the railways, first above but also later on below ground as well, and their little sisters, the trams, which enabled the cities and towns to grow as never before, and to spread and to sprawl to the scale of the conurbation. The familiar urban problems of the congested urban centre, the twilight substandard urban inner ring, and the spreading suburb, were all in existence before the internal combustion engine was marketed. Conditions in the inner city areas, too, were aggravated by the increase in horse drawn traffic which created a serious health hazard.

It is worth remembering, however, that although the rich usually travelled by horse and coach, the majority accepted public transport by rail, tram, or boat as the normal method of movement, while thousands continued to walk for all purposes.

Into this, in many cases, depressing urban scene came the motor vehicle in various forms. At first it seemed so much better in every way than the horse, and, like so many things, was superb in small quantities. With the introduction of the T model mass produced car the small cloud became a cloudburst. By the later 1930's it was becoming apparent in many American cities that the motor vehicle was beginning to cause such serious strains in the urban pattern that drastic changes would become necessary.

It was the extreme versatility of the automobile which made it so attractive and irresistible, first to the rich, and, owing to the startling advances of industrial technology, to the millions within a matter of a few years. American cities were the first to feel the impact, and to react in a predictable way. The world's biggest motorway system was built at rapid speed and enormous cost, connecting the majority of cities throughout the Union. But the motorways were driven ruthlessly into the cities, tearing great holes in the urban fabric. Whole city blocks were removed, using powerful compulsory purchase legislation, and some of the artefacts created, such as the great bridges, the tunnels, and traffic intersections, may well rank in history as wonders of the twentieth century. But, alas, for the most part they only served to increase the congestion in the city centres, and, as visitors and transport statisticians know, became congested with the traffic that always seems to increase to meet the motorways available.

Since any concerted environmental planning was apparently philosophically unacceptable in America, dynamic individual action took place. Whereas the move out from the city of the residential component in the form of suburban sprawl had only increased the sum total of urban congestion, the move of the other components actually reduced it. Large industrial plants were built out in the countryside, followed by shopping centres and, more recently, offices. The city had exploded, but instead of doing so in complete urban entities it had discarded individual organs in a haphazard and uncoordinated way. For many people it is now possible to park their car by, or in, the home, drive to the motorway and park their car somewhere in a sea of tarmac within a quarter of a mile of their factory or office, and, when they have time off, take their wives from home along the motorway to another vast tarmac carpark to an out-of-town shopping centre.

For space reasons I omit a description of the products they buy and the resultant pollution problems which are now worrying societies everywhere, but it is the way of life, or life style, that raises the question mark? Is this what you, the people want? Some writers about the environment seem to think so, and would like to raise the resulting settlement pattern to theoretical levels of consideration. If, however, we put forward the idea of the good life for all, remembering that the fundamental concept of the good life has not changed much over the centuries, certain criteria emerge. Everyone wants a good home for himself and his family, with adequate opportunities for obtaining food and clothing, for education right through the life cycle, for a variety of work, for cultural and recreational activity, for travel, and for all kinds of meeting places with fellow citizens.

Many traditional cities met the majority, or at least some, of these criteria, but none has met them for all their citizens.

Most cities have been reluctant to accept the restrictions and inconveniences involved in the implementation of the kind of comprehensive urban plans which planners have long been advocating, and which were so well discussed in relation to transport problems by Professor Buchanan in *Traffic in Towns*.[5] Instead, some of the larger cities and conurbations have embarked on another round of the somewhat out of date narrow approach type of road plans, although changing the terminology and using all the latest computerized techniques.

Following the powerful influence of the Chicago Area Transportation Study[6] of 1959–1962, many cities have embarked on large scale and expensive studies of a similar nature. The great danger in such studies lies in regard to their assumptions. Most, like the Chicago prototype, accept urban growth as inevitable, then carry out highly complex computerized surveys, establish trends from the facts recorded; and, following techniques made familiar by demographers, work from projections which often indicate that cities should be cut to pieces for the benefit of the motor vehicle.

Too few people seem to have read *Traffic in Towns*, which put forward a civilized and intelligent approach to the problem of transport and environment. Although many of the recommendations, such as those for precincts or environmental areas, a road hierarchy, and the maximum use of public transport, had been advanced by planners over the years, nevertheless the whole approach to traffic as an integral part of town planning was fundamentally correct. In fact, much of the problem of environmental planning is not concerned with thinking up new ideas, as with the difficulty in applying the ones which are already known.

Nevertheless, there are certain broad principles which should be observed. At each level of planning, national, regional, and local, the environmental plans should include the proposals for integrated transport systems which should not be considered separately. At each level every effort should be made to ensure an equable distribution of land uses and traffic generators. Similar objectives should be agreed in regard to environmental capacity. It is only by having an integrated system of environmental plans at all three levels that the imbalances which plague us today can be avoided. It must be accepted that the environmental and transport problems of, for instance, Greater London, or Greater Tokio, or Greater New York, or any other large city or conurbation, cannot be solved in an isolated way.

Fundamentally the traffic problem lies not so much with providing more facilities, such as motorways, roads, car parks, etc., but of ensuring an equable distribution of traffic generators, together with a really integrated approach to transport systems. It is not enough to ensure that good bus and rail services are available—they must be combined in new ways. For day after tomorrow one may visualize very different technological methods, such as light, one person hovercraft conveying people who find walking difficult, to climatically well protected collecting points, whence an urban hovercraft service, in which their personal medium can be folded and stacked, can convey them to work or other destination, with their personal means of communication available to be used if required at the other end. But for tomorrow, which is almost here, let us see some hard thinking about combined operations of bus and rail, so that bus and rail stations are combined wherever possible, and vehicles which can use both rail and road can be developed. Perhaps in many cases the existing suburban rail lines could be converted into flexible tramway systems for this purpose.

Above all, the traffic generators should be well distributed and their density strictly controlled. Every part of a city or town should be in a recognizable precinct or environmental area, with strict controls to ensure that human safety is paramount. Such things are easy to say, but difficult to accomplish. Over thirty years ago the city of Coventry started to plan its centre in the form of pedestrian precincts, and its residential areas in the form of neighbourhoods. Thirty years seems to be a long time for a few simple and obviously sensible ideas to become generally accepted, let alone to be acted upon. Fortunately, in the case of the urban environment, it is never too late to start.

REFERENCES

1. Thomas Adams, *Outline of Town and City Planning*. Philadelphia (1935).
2. *The Living Bible*. Tyndale House Publishers (1971).
3. Italia Nostra, *Venice for Modern Man*. Venice (1962).
4. The City and Royal Burgh of Edinburgh, *Development Plan Review*. Edinburgh (1966).
5. C. D. Buchanan *et al.*, *Traffic in Towns*. London (1963).
6. J. Douglas Carroll *et al.*, *Chicago Area Transportation Study*. Vols. 1, 2 and 3. Chicago (1959, 1960, 1962).

ENVIRONMENT AND THE TRANSPORT PLANNER

P. K. McILROY

Freeman, Fox and Associates, London and Edinburgh, Edinburgh EH3 7QJ, U.K.

The transport planner sees the problem of the conflict between the motor car and the environment in today's conditions and foresees how the problems are going to grow in the future. He strives to provide the means to satisfy the growing demand for road space but he is conscious that his solutions can be, in themselves, damaging to some aspects of the quality of life. He faces the dilemma of deciding where to strike the balance between these conflicting considerations. The planner has to suggest this point of balance; in order to do so, he must be able to make logical and credible forecasts of the results of different kinds of transport policy and he must be able to present these clearly to the politicians and the public. So that he can produce such forecasts, he has constantly to improve and refine his techniques; the paper indicates some of the ways in which this is being done in Great Britain today.

INTRODUCTION

The Transport Planner

Men have been transport planners for centuries. Cart tracks, military roads, strategic highways and bridges, railways, canals, autobahns and now motorways—these have all resulted from man's drive to expand, to open up new possibilities, to develop. The dramatic rise in population in recent centuries and the advent of the cheap car have combined to present some frightening problems, but the underlying aim of the planner remains that of making travel easier, faster and more efficient.

Until comparatively recently the enterprise of transport planners has been praised and applauded almost universally; devices to speed traffic and ease movement have managed to keep pace with the increase in demand and the job of the transport planner has been to invent, introduce and maintain such devices. Traffic flows in Britain's cities have more than doubled in the past 20 years but traffic management, one-way street systems, signal controls and improved driving standards have made it possible in most cities to cope with the resulting flows.

The Transport Problem Ahead

Although there are some signs that population increase is now being curbed there is no abatement in the rate of increase of car ownership and there is no reason to expect anything other than a further doubling of traffic flows during the next 20 years. In most cities, main traffic routes pass through busy shopping areas both in the centre and in the suburbs

and provide direct access to shops, offices, factories, filling stations, hotels and houses. This has already led to conflicts between the requirements of moving traffic and of those who live and work in the areas through which the traffic passes. It is perhaps ironic that the original reason for building premises alongside main traffic routes was because of the good accessibility afforded by the roads; to an ever increasing extent occupants of such buildings now find their accessibility reduced because of the requirements of others. These conflicts have also had their effect on the operation of buses, which, with the decline of suburban railway and tram systems, have become the most significant form of public transport in our cities. Bus speeds have been affected badly by congestion and this has tended to destroy the reliability of the service. This aspect of the transport problem is bound to get worse unless special and drastic measures are taken to prevent this happening.

The Dilemma of the Urban Environment

The transport planner sees the problem today and foresees how it is going to worsen in the future. He strives to provide the means to satisfy the growing demand, but all the time he is conscious that his solutions can be, in themselves, damaging to some aspects of the quality of life. The great dilemma is to know where to strike the balance between these conflicting considerations. The planner has to suggest this point of balance: in order to do so he must be able to make logical and credible forecasts of the results of different kinds of transport policy and he must be able to present these clearly to the politicians and the public.

THE FUTURE DEMAND

The Basis of the Prediction

For a good many years now it has been realized that there is a direct relationship between trip-making by the members of a household and the socio-economic status of the household. Clearly, the more people in the household the more trips that are made; in particular, the more workers, the more work trips. Even more significantly, it has been shown that households with high total incomes make more trips per day than households with low incomes; while with all other things equal, the mere fact of the possession of a motor car results in about two more trips being made on average each weekday by the household.

Figure 1 shows with dramatic clarity the effect of increasing income and increasing car ownership on the number of trips made per day from a household with one employed resident. Statistics on household structure and car ownership can be obtained from the National Census and surveys can ascertain effective household income. Thus, if the transport planner knows the population of each part of his study area and a few simple facts about the structure of the population, he can construct a model which uses these facts to produce estimates of the number and type of trips emanating daily from each part of the area. Furthermore, by forecasting changes in population, income and car ownership into the future the transport planner can use his model to give him a good basis for predicting the amount of travel likely to occur in the future.

The Scale of the Forecast Increase in Travel

Current predictions of the increase in car ownership

are illustrated in Figure 2 which shows that there is likely to continue to be a steady growth at least until the mid-1980's when the growth rate will, presumably, begin to tail off as the great majority of households acquire one or more cars. A common

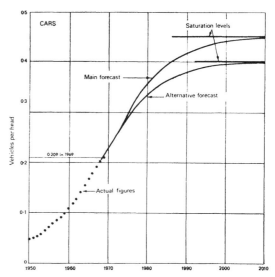

FIGURE 2   Actual numbers of vehicles per head, Great Britain 1950–1968, and forecasts for 1969–2010.
Source: LR288—Forecast of vehicle and traffic in Great Britain; 1969, by A. H. Tulpule, Road Research Laboratory.

prediction of transport studies is that the amount of traffic in a typical city in Britain is likely to double between now and 1991. The corresponding decline in public transport usage is likely to be less marked, partly because there are large sections of the community who, by reason of age, infirmity or even disqualification, cannot use cars; a typical prediction is for a drop of about one-fifth to one-quarter in public transport patronage during the next 20 years.

WAYS OF PROVIDING FOR THE DEMAND

Roads

As stated in the Introduction, the transport planner tends first to look for ways of satisfying demand. With a huge increase in the demand for car and commerical vehicle travel and with a slight decline in public transport he is liable to look first for the possible effectiveness of the present street system in trying to cope with the demand. Traffic management on existing streets—the use of linked traffic

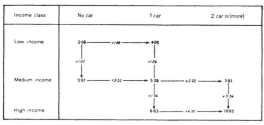

FIGURE 1   Trip making, income and car ownership.
Trip generation rates per household on an average 24 h weekday in 1962 for households with one employed resident in areas of medium bus and rail accessibility.
Source: London Traffic Survey Vol. 2, Table 14-7.

signals, the introduction of one-way streets, the removal of bottlenecks in the system—can go a long way to improving present-day conditions for drivers in cities. However, they sometimes result in worse conditions for other users of the highway, including pedestrians and buses.

More positive measures are generally needed and these invariably take the form of new roads and parking structures. In many cities in Britain the potential value of traffic management is now nearly exhausted and the introduction of new roads has begun. Naturally, with road safety at the top of his list of criteria for road building, the transport planner tends to suggest that new roads should be separated from existing roads, usually by elevating or depressing the new road; for the same basic reason, he is also anxious that new roads should, generally, be dual-carriageway and should have adequate width to allow for smooth, safe driving, leaving spare capacity in case of accidents. The result of applying these sensible criteria is the much-derided urban motorway or expressway and the multi-storey parking garage.

Public Transport

On the other side of the picture, the transport planner is concerned to maintain a satisfactory basic public transport service. Falling patronage combined with increasing road congestion contributes to the continuing decline in quality of service provided by buses and railways, and in the face of this it is difficult to put forward revolutionary and elaborate plans for future public transport in our cities. Nevertheless, the practical transport planner has some tools to help him; the quite recent change in Government attitude to the future of public transport now makes it possible to propose the construction of special roads for buses alone and the restriction of other kinds of traffic on some city roads in order to give the buses priority rights-of-way through hitherto congested areas. Measures such as these, combined with improvements in the design and frequency of bus services, are commonly put forward in transport plans and are actually being implemented already in many British cities.

Thus the transport planner is able to put forward a variety of measures designed to provide for the changes in travel demand in the future.

ENVIRONMENTAL PROBLEMS AND CONSTRAINTS

One of the fundamental difficulties of transport planning is that, while the future problem is clear and while a variety of means exist which, if applied, could solve the problem, the solutions in themselves are liable to cause distress in one way or another. The term "environment" embraces a wide range of concepts which can loosely be termed "conditions for living". Clearly, traffic and the roads and railways on which the traffic moves affect conditions for living. These conditions cannot readily be measured but it can be said that an increase in traffic on existing streets will worsen conditions for those who live, work or walk on those streets by virtue of increased noise and increased physical danger to pedestrians. Other detrimental effects of increased traffic on an existing street are a worsening of the present level of visual intrusion by vehicles and, by making it more difficult to cross the road, an increase in the degree of severance of activities on one side of the road from those on the other.

Conversely, a reduction in traffic on an existing street can be said to improve environmental conditions along that street; in order to achieve such a reduction in the face of the huge increase in demand for road space described earlier, the transport planner must provide somewhere else for the traffic to flow. The introduction of a high-class new road in an urban area can have a drastic effect on the character and environment of the area. The new road may involve the demolition of housing and the disruption, to a greater or lesser degree, of existing communities which are notoriously difficult to define until the threat of a new road brings them suddenly to light. Both during the construction of a new road and after its completion, it can cause extra noise, severance and visual intrusion in the area through which it passes. Thus the environmental planner has no easy task in setting objectives and constraints to guide the transport planner in the selection of routes for new roads and measures for improving existing roads. However, a very careful and detailed analysis of a city will generally reveal "cracks" in the cohesion of the city and, provided all the factors mentioned above are weighed in the balance, it is usually possible to postulate certain lines of opportunity.

THE BALANCING PROCESS

The original objective of the transport planner, designing roads and railways to cope with demand, must therefore be amended; however, it need not cease to be a positive objective and the transport planner must strive both to cope with the demand and to improve the environment. The process includes deciding which existing streets should be relieved of the effect of traffic, realising the presence of certain lines of opportunity and then, acting within these constraints, proposing a range of possible design solutions. These can take the form of different kinds of road system, different kinds of public transport and different kinds of philosophy on the extent to which the full future demand is to be satisfied or restrained.

This last point is becoming increasingly important as the difficulties of urban road construction come to be realised. In spite of the huge increase in car ownership, it is reasonable to suppose that, if the majority of voters decide that they do not want to see substantial new road construction near the centres of their cities, then they will vote for policies which, by parking restraint or charging, will restrict their freedom to move about in their own cars on the roads in their cities. What is not so clear is what will happen to potential car drivers in the future who will, under these circumstances, not be able to make their trips by car. It is easy to imagine that those who do not own cars today could be inhibited from driving in cars to the city centre when, in the future, they acquire cars. It is not so easy to imagine people who drive at present taking kindly to being forced out of their cars into buses or trains, with all the well-known obvious disadvantages to the individual entailed by this.

In fairness to the public, the transport planner should consider these possibilities and should put forward a number of alternative schemes each containing the appropriate policy on road-building, public transport, restraint and car parking. These schemes should all be subjected to "tests" of the future demand, using the transport model described earlier. The test of each scheme will show how trips are produced, and how they are made under each set of combined circumstances. The transport planner can show, from the results of the tests, how each scheme would work in practice in terms of passengers carried by bus services, cars and commercial vehicles travelling on roads, time taken to make each individual journey by different modes and many other related factors. He can estimate the cost of providing each scheme in relation to the likely benefits to the users of the scheme, making assumptions on such things as the value of a man's time. By this means he can tell whether the extra expenditure on roads in one scheme as compared with another is matched by a commensurate

increase in travel benefits and can, from this sort of analysis, find out which scheme is the best buy.

In addition, he can make an estimate of the effect of each scheme on the environment in terms both of increases or reductions in traffic and consequent traffic noise, etc., on existing streets; and the effects of new roads and railways and of the traffic using them on the areas through which they pass.

There are other matters which are related directly to transport; these include the effect of new roads and of different degrees of restraint on the use of the car on the commercial viability of, for instance, shopping areas; it seems fairly plain that, to an increasing degree, the viability of shops will depend upon adequate access by car and it may be that the restriction of the use of the car for environmental reasons can result in the decline of the shopping area which is itself an essential component of the environment.

The choice from among the range of schemes tested is thus a very difficult one involving judgment at several different levels. The transport planner himself has the advantage over the layman of at least some training and experience in judging between the relative importance of the conflicting factors in the choice. On the other hand the layman is frequently able to cut right across the technical and professional impartiality of the transport planner and state with startling clarity that his priority is that his own home shall not be affected by any transport proposals no matter how serious the future problem is likely to be.

THE TRANSPORT MODEL—ADVANCES IN TECHNIQUE

General

Because of the increasing demand for accuracy in forecasts, the transport planner has had to make rapid and significant advances in technique in his attempts to forecast future travel demand and to predict how this will be carried by roads, buses and railways. Figure 3 shows, in generalized form, the arrangement of a typical transport model designed for the study of a city in Britain with a population of 0·5 m., a study in which all the problems outlined in this paper played a part. As can be seen from the diagram, the model is required to produce estimates of trips by different modes of travel on an average weekday; these are then split into peak and off-peak trips and, after consideration of speeds, flows,

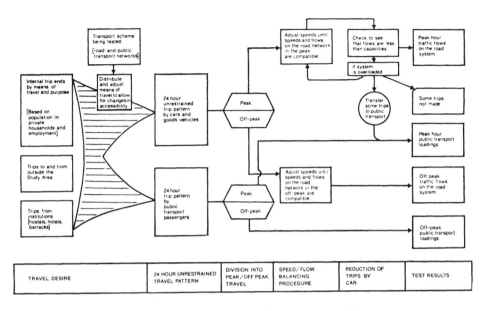

FIGURE 3   General form of the transport model.

road and parking capacity restraints and trip diversion from one mode to another, the model produces estimates of peak hour traffic flow on the road and public transport system and corresponding flows for a typical off-peak hour.

Trip End Estimation

The way in which trips are forecast has already been described in the paper. In the earliest days of the development of the art of forecasting a great many factors were thought to have a bearing on travel habits and massive surveys were taken to find out how significant each of these was. As a direct result of this pioneering work, mainly in London and the West Midlands, it was found possible to produce realistic estimates on the basis of comparatively few socio-economic factors and these now form the basis of what is known as the category analysis. Average trip rates have been established for each of 108 different types of household and it is now necessary merely to carry out small sample surveys in order to reflect local variations from the average in order to produce convincing results for any study area.

Trip destinations are generally forecast on the basis of an analysis of the amount and location of employment. For example, an employee in a retail store may serve 25 customers a day; each of these

25 people has made a trip to that employee in that store.

Multiple Route Assignment

The majority of drivers who are familiar with their city and make regular journeys, for instance to and from work, tend to use the shortest route available to them. However, the introduction of new roads and of traffic management measures, combined with the congestion effects of increased traffic, make this generalization rather dangerous. Even without these complicating factors, there are always those who for one reason or another will choose to travel by different, less convenient or slower routes. Observation has shown that it is possible to estimate the statistical probability of the choice between the fastest route, the next fastest and so on and this concept has how been introduced to the assignment process available to the transport planner.

A further refinement which is becoming increasingly important as new, fast roads are contemplated in our cities, is required to reflect an additional consideration affecting the choice of route. Whilst the introduction of a by-pass to a congested area can result in faster trips, the trips tend to be longer in terms of distance and therefore more costly to the owner of the car. In addition, the cost of car parking and other charges in city centres can have a further effect on both the choice of route and the choice of mode of travel. Thus the concept of "generalized cost" which includes elements of time, running cost and parking charges is beginning to supersede the use of time alone as the main factor in the choice of route and mode.

Public Transport Assignment

The prediction of car trips and their assignment to road systems is complicated enough, but the application of these principles to public transport—say a bus system—is even more of a problem. The model has to be able to take account of the routing and frequency of bus services, the time taken to walk from the origin of the trip to the bus stop, the wait for the bus, the reluctance of passengers to change from one bus to another during a journey and, finally, the walk from the bus to the destination. The analysis of all these factors is somewhat cumbersome and offers enormous scope for development of the technique. A good deal of research is being undertaken in this field.

Peak and Off-Peak Conditions

The transport planner who is concerned with the actual design of urban roads is chiefly interested in the likely conditions of heaviest loadings, typically occurring in the morning and evening peak hours. Conversely, the environmental planner tends to be more concerned with traffic conditions during the day in shopping areas and during the night in residential areas. Thus the model must distinguish between peak and off-peak conditions and this involves defining a peak hour and a typical off-peak hour and then isolating trips made for different purposes during those hours from the all-day pattern of trips. Further than that, the transport planner must make a series of assumptions about the possible effects of the staggering of working hours and other changes which may alter the proportion of trips made for certain purposes during the peak hour. Estimates of these changes are necessarily somewhat speculative and serve to highlight the fact, that, however intricate the model, the quality of the results depends upon the thinking of the transport planner in providing the information for the computer.

Speeds and Flows on the Road

There is a tendency for average speeds to fall as the amount of traffic increases and it has been found possible to measure the relationship between speeds and flows on individual sections of road. Figure 4 shows the general form of relationship, which becomes unstable as flows approach capacity of the road. Factors influencing the shape of this curve include the width of the road, the distance between junctions, the method of control at junctions and the general character of the road surface. The assignment process has to take all these factors into account by a series of adjustments which allow the system to converge gradually to a point of balance in which speeds and flows are compatible throughout the network.

Restraint on the use of Cars

In attempting to simulate the effect of policies of restraint within the model the transport planner has to distinguish between different kinds of trip and different methods of restraint. Thus, in a scheme which provides very little new road building there will be too much traffic for the road system and a car parking policy will have to be introduced to

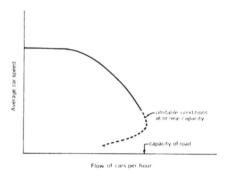

FIGURE 4   Speeds and flows of traffic.
The curve shows the typical form of the relationship between speeds and flows on a road.

throttle down the demand to a level which can be accommodated by the road system. Clearly such a policy can be applied with sufficient sensitivity to allow the maximum possible number of trips to take place by car and a linear programming technique has been developed to help in achieving the optimum policy.

Such a policy should, of course, be capable of reflecting the desire to prevent damage being done unnecessarily to the commerical viability of the city centre so that short-stay parking should be treated differently from long-stay, which is predominantly used by trips to and from work. These considerations have a very strong bearing on the design of the model; again, the transport planner has in his hands a powerful tool but he must be fully aware of the implications of the process which he applies in his study if he is to make sense of the results of the complex mathematical procedure.

CONCLUSION

This paper has set out in very broad outline the problems faced by the transport planner, the environmental considerations which increasingly constrain his search for solutions and the way in which his techniques have evolved during the past decade to allow him to analyse the situation and to predict the likely effects of a variety of policies.

The problems are clear enough but the possible solutions are many-sided and frequently there is conflict between the need to solve the traffic problem and the desire to preserve environmental quality from the effects of new roads. This is a field in which there are no absolute values and in which criteria vary from person to person. Often the motive in objecting to a proposed new road is simply a natural resistance to change; as often it is due to fear that the new proposal might affect the value of the property of the objector.

It is the task of the transport planner to demonstrate all the advantages and disadvantages of various courses of positive action and incidentally, of doing nothing whatever about the problem, but the final decision remains a political one in which the strength of public feeling about each of the factors must be gauged.

ENVIRONMENT AND THE NEW MOBILITY

JAMES ALAN PROUDLOVE

Civic Design Department, The University of Liverpool, P.O.B. 147
Liverpool L69 3BX, U.K.

Increasing private ownership of cars continues to create increasing pollution, noise and congestion from traffic. A worsening of the urban environment results, especially in areas of high traffic density. The car is, however, filling an increasingly important role in contemporary society, permitting the freer exercise of choice, greater variety and more fulfilment in life. It may even be the most economical transport system. Examination shows that the environmental objections to traffic and the car could be largely removed by an improved design of vehicle, by improved town design, and by a greater responsibility amongst drivers. These measures could be an effective alternative to traffic limitation as a means of safeguarding or even improving the urban environment.

INTRODUCTION—TRAFFIC AND THE ENVIRONMENT

That the environment could be improved is commonly agreed, although it is doubtful whether many of the familiar remedies result in improved social, as opposed to sanitary, conditions. It may also be true that the environment is worsening. There are increasing sources of industrial air and water pollution, although many of the worst excesses of early production have been removed. Similarly air pollution from domestic heating must be incomparably less than it was seventy years ago, and last century's descriptions of river filth suggest that river conditions then were at least as bad as today.

Almost certainly we are much more aware of environment and pollution than ever before, no doubt as a result of the efforts of a few devoted environmentalists. Their warnings have been taken up by a public which is beginning to be able to afford the luxury of a good environment.

The particular aspect of the problem which concerns us here is that of the built urban environment and its faults, at least those aspects which give rise to much of the stress of urban life: congestion and the increasing difficulty of travel; and the danger, noise, smell, and general intrusiveness of traffic into everyday life. It has become fashionable to equate traffic problems with the motor vehicle *per se*, whereas careful analysis reveals more fundamental causes.

The case against the motor vehicle, and particularly the car, has been well rehearsed of late, and it is undeniable that in certain respects the car is an intolerable nuisance. Its power unit creates air and water pollution, noise and smell. The vehicle intrudes into our lives and creates noise, obstruction and sometimes danger.

Society's current love/hate for the car has been fostered by the xenophobic view of some conservationists which has been presented by the mass media as the contemporary view. Yet the question which remains unasked is whether the problem lies with the car *per se* or with the way in which it is designed and used.

Most of the objections fall into one of two groups, the first concerned with environmental pollution, the other with the physical presence of vehicles. Noise, that particularly topical problem, can be included in each group whether it is regarded as an area-wide problem or, alternatively, as a very local one.

It is in towns that the problems caused by the car become acute and the more so at the higher densities of development. In rural areas the noise of a car quickly diffuses in all directions and is absorbed by the soft surfaces of earth and vegetation. In towns, on the other hand, streams of cars create a continuous noise source which is contained between buildings and reflected by the hard surfaces of walls and street to intensify the loudness in that vicinity. Acceleration noise near junctions, bus-stops, or corners, occurs more frequently in town than in country, and fuel consumption per mile is also higher, resulting in more emission and greater atmospheric pollution, less easily dispersed due to the enclosure of the streets by buildings.

The air pollution of southern California was perhaps the first to be widely recognised. There two effects combined to create the particularly obnoxious condition of the Los Angeles "smog". California has the world's highest level of car ownership and probably also the highest average annual mileage and gasoline consumption per head. When coupled with the peculiar micro-climate, the photo-synthesis of exhaust emissions produces the famous "smog" containing aldehydes, which is held down near to the ground by the temperature inversion common to that region.

Elsewhere the popular outcry against traffic-generated air pollution is of quite recent origin, as is the concern about traffic noise. Both have come to the fore as a result of the rapid increase in the number of motor vehicles, but also because of increased public awareness aroused by the media. The danger to health from pollution and noise has been established as real where either occurs above certain thresholds, but physical danger is also ever present from the proximity, and sometimes the mingling of vehicles and pedestrians.

A quite different and more sophisticated case against the car as an environmental detractor is that which complains that there are just too many cars around and that they obtrude to an unacceptable degree. Such a view must of course be highly subjective and dependent upon the individual standpoint. To the motorist too many other cars cause congestion and delays, a restriction of individual choice of speed and action, and probably the inability to find a convenient parking place. But it is always the other car rather than one's own that causes the congestion, "they" are causing "us" problems. But generally speaking the motorist can reach almost every place in a town because it is conventional to provide roads for vehicular access to every property. In the newer parts of towns "bylaw streets" have ensured that they are, in most cases, of sufficient size to accommodate the demands of even today's traffic, except where development densities are high and no provision is made for vehicles to be stored off-the-street.

Such, of course, is not true of the older parts of towns, built before the current era and often being redeveloped at ever higher densities. Streets there will have been resurfaced and marginally improved to deal with modern traffic. Usually the demands upon them result in their being devoted to moving traffic, with the accompanying traffic management orders and equipment to secure this objective. Again it is the high intensity of traffic which creates the conditions which disturb. Highly managed traffic flows such as in one-way streets, apart from

creating problems of noise and pollution, create serious barriers to movement between adjoining districts, leading to danger where pedestrians must cross and loss of harmony within areas through which the traffic passes. It is here that the car can, with justification, be blamed for a situation which must often be regarded as intolerable. But even this conclusion must be tempered when one remembers old prints showing these same streets bursting with equally frantic activity a century ago. Almost certainly there has been a radical change, a sharpening of public attitudes towards traffic, as more and more streets become subjected to high levels of traffic use. It is the mixing of disparate activities, broad new traffic routes, and the presence of vehicles close to buildings, which most worry the conservationists. Yet there must always have been a good deal of pedestrian and vehicular activity, particularly in town centres. It is this activity which brings an area to life. Undoubtedly it is the large volumes and the thrusting nature of traffic which cause most public resentment even though, above a certain level, increased traffic brings little increase in noise or danger to pedestrians.

A social case has also been made out against the car. A marked decline in public transport services is the corollary of a high level of private car ownership. The decline has left the residual, non-car-owning section of the public at a disadvantage in terms of mobility and opportunity, particularly for access to work and recreation. Most of this group are likely to belong to the low-income "under-privileged" end of society, whose problems will thereby be increased. By restricting the ownership or use of private cars, it is argued, the greater demand for public transport so created will maintain its economic viability and ensure a high quality of public transport service.

THE CAR IN SOCIETY

Whilst the private car first appeared in towns some 70 or 80 years ago, in minute numbers at first, vehicular traffic was already a problem. Serious vehicular congestion at the centres of the larger towns in Britain led to a Commission which reported in 1910 on the need for by-passes.

The present situation, and that which can be clearly foreseen, is very different. Ownership of private cars has increased steadily through the twentieth century and, except during wars or severe depressions, their number has doubled about once a decade. During the early years of the century this increase must have replaced other forms of private transport, to the virtual extinction of the horse drawn vehicle by mid-century. Latterly the ownership and use of private cars has extended more widely to cover almost the whole spectrum of society in industrialised countries. The rate of increase in the number of private cars is now expected to decline, to produce about double the present number, if the population remains static, as it approaches the level at which it is believed the market will become saturated.

Vital to contemplation of the future of the motor car is an understanding of its role in society now and in the future. Until the last decade this role was unresearched but recent studies, such as the London Traffic Survey,[1] and the Leicester, Reading, West Midlands,[2] and many other studies, have begun to reveal many of the behavioural habits of town dwellers and of the different life styles of car-owning and non-car-owning families. The car owner has also been studied as a user of public transport, to throw light onto the mechanism of choice of travel mode for those for whom either mode is available.

Such studies have revealed, not surprisingly, that more richer families tend to belong to the car-owning categories, and that the latter, whether or not rich, travel more extensively than their non-car-owning counterparts. This difference is important and fundamental to the role of the car in society because it reveals the richer variety of life style, where this is measured by mobility and opportunity, of the car-owning sector of our society. The difference is not the simple result of affluence alone but is similarly exhibited by the car-owning category amongst the lower income groups.

Particularly significant are the types of activity which car availability bestows. There is little difference in the number of journeys made to work-places by car-owning or non-car-owning families, but the former travel very significantly more than the non-car-owning for all other non-work trips. Thus whilst the car perhaps allows journeys to work to be made by private rather than public transport, the number of journeys to work will be little changed as a result of increasing car ownership. However what will change will be the extent of travel for leisure activities, which may be taken to include shopping trips. Presumably this type of trip will often be made by several members of the family together. The converse of this picture is

also significant, in that many trips fewer would be made if car travel was not permitted, resulting in an overall lowering of participation in many fields of activity.

Higher levels of car ownership are likely to lead to more car-pooling for work journeys and a greater availability of cars for other members of the family. We are probably also experiencing the emergence of a new sphere of usefulness for the car as more women become drivers and proportionately fewer cars are used for the journey to work. As a result the housewife is less tied to the home and becomes better able to participate in her chosen "extra-mural" activities. Development trends too are leading to the car being more and more necessary both for basic activities and to participate in "new" activities such as outdoor recreation, discount or one-stop shopping. The recent trend towards the spread of employment, shopping and entertainment facilities throughout the suburbs means that, without a car, those places which are not in the home sector are relatively inaccessible, thus limiting choice of workplace, shopping centre, cinema, etc. Such a situation is the opposite of that created by the advertising media which reveals to all the breadth of choice available only to some. Greater satisfaction can be seen to be an important justification for car ownership.

As the car becomes more widely adopted as an item of necessity for each household, the cost of car travel comes to be regarded simply as the perceived cost of a particular journey, often as no more than the cost of parking. The cost of fuel, which is about one-quarter of the total running cost, will usually be less than the cost of most public transport. The true cost of private car travel, to an individual, is well above the price by public transport, but the difference quickly disappears when passengers are carried in the car.

Cost differences, to those who have the choice of either public or private transport, have been shown to be of relatively little significance. Car owners' valuation of speed and convenience, riding comfort and companionship, are evidently highly rated. Experiments, such as free public transport, have limited success in attracting car owners but are likely to show increased usage by those already dependent upon that means of travel. This is perhaps a desirable objective in itself, reducing the differential mobility of the two groups.

Society has clearly adopted the car for the travel freedoms which it confers on its users: to choose to go their own way; for the man who values his time and comfort highly, and perhaps even for real financial gain over the cost of public transport for family travel. Whether or not this freedom should be subjugated for the good of society is the subject of contemporary debate. Most of the argument heard suggests that private car users are anti-social by creating danger and pollution, or by depriving others of services or environmental quality. Reason requires, however, that the basic problem be examined rather than that the effect should be attributed mistakenly to the wrong cause.

In considering any limitation on the use of the private car it is also essential to keep in mind its advantages, now so widely available, to ensure that the societal gains resulting from today's mobility[3] are not needlessly sacrificed to achieve some undefined environmental advantage.

A POSITIVE APPROACH TO ENVIRONMENTAL IMPROVEMENT THROUGH DESIGN

In the previous parts of this paper it has been shown that, whilst the use of increasing numbers of cars in towns is creating serious environmental problems in many areas, the vehicles are fulfilling a real societal need. Improvement of the environment based on a reduction of vehicle use would be contrary to popular wishes, and represent a misunderstanding both of the need for private vehicles and of the root causes of the problem, which arise from poor vehicle design, intolerance in the use of vehicles by some drivers, and unsuitable town design.

The environmental objections to the car can be resolved into two groups: those which relate to vehicle performance, i.e. its appearance and its emissions, with which noise can be included; and those which relate to its use, particularly danger and intrusiveness. Resolution of the first group of objections is then a matter of vehicle design, particularly insofar as this is influenced by choice of power unit, whilst the second is a problem of layout of town or country, and of human behavioural characteristics, particularly those of vehicle drivers.

Although reportedly intense efforts appear to have produced little change in the conventional internal-combustion engined vehicle, there is little doubt that a vehicle could be manufactured to be near pollution free. Most of the contemporary effort appears to be devoted to "cleaning-up" the present generation of vehicle, by better sound muffling and more complete fuel combustion.

Further effort is required to develop a vehicle which, intrinsically, is quiet and pollution free. There is at present little incentive towards this end whilst the motor industry occupies such a prestigious position and the private car has become so widely available. A specification for such a vehicle might require that its motive unit produced only hydrogen and oxygen wastes and that its maximum loudness should not exceed say 60 dB outside an envelope 3 m from the vehicle. Perhaps little fundamental progress in this direction will be made whilst the world supply of hydrocarbons continues to grow!

Hopefully the initiative established by the State of California will be followed by other industrialised nations. This should at least lead to "cleaner" internal-combustion engined vehicles and progressively more restrictive control on the emission of noise, and consequently to a more acceptable, if not blameless, vehicle.

Turning from the universal problem of air pollution to a review of the nature of environmental disturbances attributable to the use of motor vehicles, these can be classified as arising from either a high density of vehicles, or the intrusion of even single vehicles into particular areas of high environmental quality.

High densities of vehicular traffic give rise to noise and danger in busy streets, to the segregation of activities within otherwise cohesive areas, and to the submergence of some areas beneath a tide of parked vehicles. In most towns the busiest traffic routes tend also to be shopping streets and the location of most of the town's busiest activities. This apparent conflict of interest has historical origins which are fundamentally logical. The centre of activity is undisputedly the best location for a public transport serving these activities. It is only with the growth of private vehicles that most of the traffic has become extraneous to the streets which it traverses. The presence of this traffic in busy activity centres and the frustrations, on both sides, which its passage creates, lies at the heart of the antagonism towards the car *en masse*.

Large concentrations of traffic occur at particular parts of the road network for a number of reasons. The first will be due to high levels of traffic generated by adjoining development, but this effect may be extremely local. Or a road might attract much traffic to itself because it is an *arterial*, in the sense that it is broad and has a large traffic capacity, but also because it *articulates* well within the total road network, fulfilling the directional needs of major traffic flows.

This suggests two ways in which design could be employed to reduce the effects of vehicular traffic concentrations. First, excessive concentrations might be avoided by suitable road layout to spread traffic more uniformly over the network, Secondly, the effect of such concentrations as must occur can be mitigated by ensuring that the main traffic routes are located around (rather than through) environmentally cohesive areas such as shopping streets or housing estates.

The geometry of the whole network is itself an important consideration affecting traffic loading. Most journeys using a ring/radial pattern of streets are drawn to the innermost ring because this is the shortest distance route across town. A grid network, on the other hand, tends to equalise traffic density by providing many paths of similar length between points in the town and thus avoiding central congestion. A grid provides good route articulation but diagonals[4] or high grade inner-ring-roads must be avoided because of their effect in concentrating traffic flows.

Pursuit of these objectives might lead to the criticism that the result would be traffic spread more thinly but over more of the town than would otherwise be the case. Such a point of view inevitably leads to the question "in what circumstances would moderate volumes of traffic be acceptable". Unless one adopts the arbitrary posture which demands that vehicular traffic must be stopped from entering towns, some equitable arrangement must be achieved.

There is a long history of attempts to introduce precinctal or environmental area planning into town design, but still many plans are produced in apparent ignorance of these principles. The case for traffic routes lying outside environmental areas was made particularly well in the Buchanan Report *Traffic in Towns*,[5] and plans for New Towns[6,7] produced since its publication usually incorporate its principles. The principle is much more difficult to apply to old towns, but in many urban renewal is providing the opportunity to introduce these ideas. The difficulties are amply illustrated in the reports on Bath[8] and Edinburgh[9] prepared by the Buchanan team.

Other measures would include avoidance of the creation of large generators of traffic concentrations, such as big industrial estates with high worker densities and peaks of arrivals and departures.

Vehicles *en masse* are a common enough hazard in many parts of most towns, and affect most people there in one way or another, such as danger to the pedestrian and separation of districts, congestion to the driver, delay to the bus passenger, and an ever present unwelcome neighbour of many housing areas.

At a different scale even single vehicles can create intense disturbance when they penetrate into areas which are essentially quiet or still or visually complete. In such areas it is justifiable, within reason, to consider severe restriction on the use of motor vehicles in the cause of environmental protection but, here again, much can be achieved by design to secure these aims. All buildings require some immediate vehicle access for servicing, but few generate such a level of traffic as would lead to problems resulting from high volumes. However, a more disruptive effect is created by a single servicing vehicle in an environment with a low ambient noise level than would the same vehicle servicing a building sited on a busy street. Whilst design can achieve much in these circumstances, such as by siting roads as far as possible from buildings and avoiding the need for difficult or restrictive vehicle manoeuvres, which lead to excessive noise, there remains the need for consideration on the part of vehicle operators. Recognition must be engendered that in some places the vehicle is subjugal to the environment. This must be accepted as a condition of entry to them. This is the fundamental principle of environmental area design.

Design improvements are usually achieved, unfortunately, at a considerably greater cost than the *status quo*, especially when they are assessed sectorally. Increasing awareness of the environment will no doubt lead to the evaluation of good environment, when its costs can be incorporated into the overall cost-benefit exercise. Then a different picture may emerge. The overlooked theme of *Traffic in Towns*,[5] which is becoming increasingly applicable, is the Buchanan Law that ENVIRONMENT = COST/TRAFFIC VOLUME. This law, which equates increasing traffic to declining environmental quality, in the absence of expenditure, clearly relates the three variables of the design process. It also shows that good design can be an effective alternative to traffic restriction for the achievement of a good environment. Only by this approach can we secure a good environment without the sacrifice of the new mobility.

REFERENCES

1. Freeman, Fox, Wilbur Smith and Associates, *London Traffic Survey*. L.C.C., London (1964).
2. Freeman, Fox, Wilbur Smith and Associates, *West Midlands Transport Study*. Birmingham (1968).
3. C. D. Buchanan, *Mixed Blessing: The Motor in Britain*. Hill, London (1958).
4. H. T. Fisher and N. A. Boukidis, "The consequences of obliquity in arterial systems" *Traffic Quarterly* **17**(1), 145–170 (1963).
5. C. D. Buchanan, *Traffic in Towns*. H.M.S.O., London (1963).
6. Shankland, Cox and Associates, *Ipswich: Draft Basic Plan*. H.M.S.O., London (1968).
7. Llewelyn-Davies, Weekes, Forestier-Walker and Bor, *The Plan for Milton Keynes*. MK Development Corpn. (1970).
8. C. Buchanan and Partners, *Bath: A Planning and Transport Study*. London (1965).
9. C. Buchanan and Partners, *Analysis of the Problem: City of Edinburgh Planning and Transport Study*. Edinburgh (1971).

ECONOMICS, TRANSPORT PLANNING AND THE ENVIRONMENT

K. M. GWILLIAM

Institute for Transport Studies, University of Leeds, Leeds 2, U.K.

Environmental quality is recognized as a proper objective of transport planning presenting extremely difficult technical design problems. But it also presents essentially economic problems in the husbanding of scarce resources and in the selection amongst alternative possibilities. The solution of these problems can best be achieved if environmental objectives can be identified, quantified, projected and evaluated with sufficient precision for them to be incorporated in a formal design criterion. Research is now being developed to this end.

There are limitations, both logical and practical, to the development of this approach. But, so long as these are properly appreciated, quantification and economic evaluation are not competitive with design skills but are proper complements to them.

THE NATURE OF THE PROBLEM

The problem of the motor car and the environment is essentially one of choice. To have more mobility we must forego some other desirable object; noise, fumes, visual intrusion, delay to pedestrian movements, community severance and so on, are amongst the opportunity costs of increased motorization. Or to put it another way, the environment may only be protected against the motor car either by having less motorization than we desire or by incurring other costs in designing our cars, our roads, or even possibly our buildings so as to limit its impact.

This is not to deny, of course, that merely by explicitly recognizing environmental objectives we may be able to design more efficiently to achieve them without increasing costs. But there is a limit to the extent to which we can put off the fundamental issue in this way. Ultimately and inescapably we must face the economic problem of allocating scarce resources between competing ends.

This does not quite get to the heart of the economic problem of the environment, however. The argument so far could be applied equally well to the production of ice cream or shoes, yet we do not consider that there is any special problem connected with these types of product. What makes the relationship between the car and the environment special as an economic problem is that it falls into two categories which have always given the greatest difficulty to economists, namely,

(i) *Externalities in production.* The motor car causes unintentional environmental costs which impinge on the rest of society.
(ii) *Jointness in consumption.* In some respects environmental quality, like national security, is a commodity which cannot sensibly be produced in discrete units for individual consumption but must be produced as a "public good" or not at all.

For these two reasons the market mechanism, which quite adequately handles the "problem" of ice-cream or shoes, does not solve the problem of the motor car and the environment.

There are well established traditional lines of approach to these two categories of difficulty. Where externalities in production arise the tradition is to "internalize" the external effects by fiscal means. Thus goods which are deemed to have adverse external effects are subject to taxes and those deemed to have beneficial external effects may be subsidized. In practice, of course, this approach depends on being able to recognize, measure, evaluate and attribute responsibility for the external effect. Even in the case of road congestion, which is probably one of the simplest externalities that can be found, we have not yet actually implemented a pricing scheme to internalize the effect. So we must expect difficulties if we are to treat the environment in this way.

With respect to public goods the decision as to how much is to be produced stems usually from concensus in the political field. For some goods, such as policing or defence, the range of disagreement about the amount to be produced may be relatively small (either because of some obvious indivisibility or discontinuity in its effectiveness or because the public are generally ignorant on the matter and will easily accept any solution). For others, and particularly for those which are borderline public goods (such as health services) which could to some extent be produced privately, there may be less consensus. The reason for this is essentially one of distribution of welfare. "Publicness", by releasing consumption from income restraints, is inevitably redistributional and as such is accepted by some and rejected by others. Both the awareness of the environmental problem, and the presently accepted approach to it, may be substantially overlaid with distributional implications.

THE PRESENT STATE OF THE ART

So much for the nature of the problem. We shall now consider the institutional and intellectual framework within which it is presently approached.

There are three distinct areas in which motorization could be controlled, namely by operating on:
(i) Vehicle design
(ii) Use of vehicles on the road
(iii) The provision and management of road capacity.

(i) *Vehicle Design*

Some disbenefits can be reduced or eliminated by construction and use regulations. Vehicle size, power/weight ratio, axle loads, noise and smoke emissions are already controlled. Whilst in some respects continental standards are more restrictive than ours (e.g. power/weight ratios), in others they are more liberal (e.g. vehicle dimensions) and British entry into E.E.C. will necessitate a thorough review of standards. The problem lies not in the willingness to exercise control, which has already been amply demonstrated, but in establishing a proper basis for control. Although there have been a number of analyses of alternative construction standards,[1] we know rather more about the resource cost implications of the alternatives than we do about the value of any environmental benefits which may result. Better information on the evaluation of environmental benefits is absolutely crucial in the formulation of construction regulations properly incorporating environmental objectives.

(ii) *Trip Restraint*

The second possibility is to restrain either number of vehicles owned or the propensity to make trips with the given vehicle stock. So far restraint on ownership has not been used as a planning instrument, though car purchasing has been manipulated by purchase tax and hire purchase controls as an instrument of short term macro-economic policy. Indeed, even where parking controls have been the main instrument of restraint policy special provisions for residents parking have usually been introduced to avoid rigorous restraint of ownership.

A quite different philosophy is beginning to develop on the question of trip restraint. Planning authorities are increasingly willing to use their traffic management powers to limit the total amount of traffic and particularly to control access to particular routes or areas. Where parking restraint appears insufficient to ensure the desired level of service for "essential" traffic planning authorities are showing interest in a range of other instruments for restraining "optional", or less essential traffic (such as area licensing or road pricing) or of supporting public transport (by capital grant, operating subsidy, bus priorities, etc.) in order to reduce private vehicle mileage. Though some extensions of powers have recently been sought (for instance in the enforcement of parking restrictions), it is not the lack of statutory powers so much as the lack of political will to use existing powers that limits the extent to which the environment is protected by traffic restraint.

This lack of will stems, at least in part, from a lack of conviction that the achievable environmental benefits are worth the loss of mobility which they cost. It also results partly from a belief that the problem can be treated more suitably by investment.

(iii) *Road Investment*

The problem of motorization in the urban environment is frequently characterized as that of incompatibility between the functions for which the roads were originally designed and the types and volumes of traffic presently using them. For example, some main traffic routes have developed commercial functions which become increasingly difficult to sustain as traffic volumes grow. Similarly residential areas may be damaged environmentally by through traffic seeking "rat-runs" to avoid congestion on the main traffic arteries.

Conventional engineering wisdom suggests a simple solution to this problem by the separation of different kinds of traffic on a hierarchically structured road network. At the apex of the hierarchy we would have limited access primary roads designed for high speed trunk movement; at its base, local roads would be completely protected from through traffic. In order to support this hierarchical structure there must be enough capacity on the primary roads to take all the traffic assigned to them and this usually implies construction of new primaries.

There are two obvious difficulties about this approach. New roads within large conurbations are extremely costly. The larger the conurbation the smaller the proportion of through traffic and the less the chance that the traffic problem can be solved by relatively less expensive peripheral by-passes. In the conurbations the strategy is bound to be expensive. In the likely event that the traffic benefits do not offer a reasonable rate of return on the investment this raises the question of the size and value of the environmental benefit achieved.

The second difficulty is perhaps even more serious. There is no certainty that, in environmental terms, the problem can be solved by increasing road capacity. New roads are difficult to fit into a densely developed urban fabric and by themselves cause adverse environmental effects. Even the advantage of channelling traffic on to purpose designed facilities can be illusory. Improved roads may generate extra trips (although there is only limited evidence of this effect) or increase average trip length (which is a well substantiated effect). Together the two effects may result in little or no net relief for the existing roads, simply redistributing the environmental pressure from roads competing with the new primaries on to those complementary to them.

The position may be summarized as follows. Improvements in vehicle technology may help but cannot solve the problem entirely. Transport planning is therefore necessary for the protection of the environment as well as for securing traffic benefits. Moreover, investment in new road capacity may not automatically solve the problem even if capital were free. Thus, implicitly or explicitly, environmental effects must be weighed one against another, and against traffic benefits foregone as well as against capital costs if a rational policy is to be formulated.

THE ENVIRONMENT AND FORMAL ECONOMIC APPRAISAL

It is increasingly common for planning authorities to specify the environment as a transport planning objective in addition to the achievement of a reasonable rate of return in a cost benefit appraisal. It is easy to see why this should be so. Environmental benefits are not presently included in the formal economic appraisals and there is no reason why they should be highly correlated with, or adequately proxied by, traffic benefits.

It can be argued, however, that environmental objectives, unlike income distribution arguments, are logically aggregatable with the other resource elements of economic appraisal. Thus, in principle at least, it is possible to conceive marginal social cost as including marginal environmental effects and to envisage environmental benefits as an additional item within a cost/benefit appraisal. If this could be achieved in practice both the design and the assessment of urban transport plans would be greatly improved. The crucial question is whether we can quantify the environmental effects in terms appropriate to this task.

The cost/benefit appraisal framework has been widely discussed[2] and needs no description here. In order to include any item of benefit or cost in such a formulation, however, it is essential that it can be

(a) clearly identified
(b) precisely measured
(c) accurately forecast
(d) objectively evaluated.

(a) *Identification*

Little need be said about the identification of dis-benefits. The items which have been most widely discussed are noise, dirt, fumes, visual intrusion, community severance, pedestrian delay, vibration, and loss of light, all of which are fairly easily identifiable physical effects.

Less tangible, and more difficult to identify, are the disbenefits arising from the expectation of future ills, which find expression in personal form as fears of various kinds and in market form as planning blight.

Considerable problems arise in identifying those losses of welfare associated not with present ills but with expected future ills.[3] Insofar as they are merely the (accurate) shadow of the future effect it would not be appropriate to include them in a formal appraisal in addition to the "real" effects themselves. We include them in our list, however, because it is likely that, certainly with respect to fear, they represent a misery which may be magnified by anticipation.

(b) *Measurement*

Ideally each of the elements which we identify should be susceptible to measurement in order that it might subsequently be evaluated and the conventions derived made to apply in the design situation. The present state of the art differs substantially in this respect. For *noise* there are too many rather than too few measures[4] and the task is to select that which is most helpful and most manageable in the design situation. Fortunately, there is substantial agreement concerning the main aspects of noise which cause annoyance (level, pitch and variability) and there is evidence to suggest that a number of measures which combine these aspects correlate very closely with each other and with subjective rankings of noise annoyance.[5] Similarly, there is some knowledge about *vibration* though it has been related to physical damage to buildings and to the road surface rather than to subjective annoyance rating.[6]

Fumes are a little more difficult. It is possible to measure, fairly precisely, the concentration of each of about 200 chemical components of the fumes emitted by motor vehicles. The difficulty is that such physical measures have not yet been correlated satisfactorily with any annoyance rating, even though there is ample evidence that some of them are physically harmful in excess.[7] Thus, strictly, the problems are those of evaluation, to which we shall return.

Though noise, fumes and vibration have been the subject of the most detailed examination more recently emphasis has shifted towards some of the other dis-benefits for which means of measurement and evaluation are being sought. *Pedestrian delay* is susceptible to an obvious aggregate time measure which does rank well with annoyance. For visual intrusion the solid angular subtense measure[8] correlates well with subjective ratings for physical intrusion, but a further variable is still required for the aesthetic valuation. For some dis-benefits no objective measures have yet been found (e.g. community severance, loss of privacy). In some. cases (e.g. the expectational problems of fear, blight, etc) it may be that the most appropriate approach by-passes the need for a physical measure

as their effects are directly related to traffic volumes or design features. In such cases the problem becomes one of direct evaluation.

(c) *Forecasting*

In order to introduce the environmental analysis effectively at the design stage there are two main requirements. Firstly, we must establish clear and stable relationships between the measured environmental effects and quantifiable, or clearly definable, design features or traffic characteristics.[9] Secondly, we must be able to forecast traffic or design characteristics with the necessary precision for the estimation of environmental costs. Work is already under way on the first problem and preliminary results suggest that useful and stable relationships can be obtained.[10]

In view of the great sophistication of the modelling procedures already used in urban transport planning, it is perhaps surprising that there can be any doubts at all on the second score. The problem can be highlighted by juxtaposing two observations. Firstly it is recognized that the assignment models commonly used are conceptually fairly crude. Where finely balanced alternatives are available traffic engineers have only limited confidence in their ability to correctly assign individual traffics. In aggregate terms, it is found that the models produce realistic loadings in particular broad channels and realistic estimates of point to point travel times and for traffic purposes this is generally sufficient. It is the fine detail, particularly on local roads, that is difficult to predict.

The second observation is that there may be a very steep threshold for some dis-benefits, with broad plateaus in traffic volume terms of low and high environmental effects.[11] Thus the environmental damage may be very sensitive to relatively small variations in traffic volume at this threshold on particular routes.

The obvious conclusion to draw from this evidence would seem to be that, in order to obtain the greatest environmental advantage from given total road capacity which might require traffic forecasting and management of a greater degree of precision, and with different objectives in view, than has hitherto been achieved

(d) *Evaluation*

The whole systematic approach to the economies of the environment ultimately depends on our ability to obtain meaningful and acceptable values, or prices, for the component characteristics of environmental quality. It is this very step which is most commonly called into question by planners.

Perhaps the most fundamental and contentious issue in this respect concerns the nature of the prices sought. The conventional economic wisdom is that we should be looking for *behavioural* prices, namely prices which represent the actual or implicit valuations which consumers put on goods or services in market or pseudo market situations. The case for this approach in the Cost Benefit appraisal context has been presented most forcefully in the Roskill Report.[12] The basic argument is that consumers observed behaviour offers the most reliable and least arbitrary estimate of the benefit which they obtain from any object of consumption.

There are, however, a range of reservations about this approach which may have relevance in the environmental context. We shall discuss them only briefly.

(i) The most common reservation concerns *income distribution*. It is correctly observed that any set of prices obtaining in the market will reflect the prevailing income distribution and prevailing technology. If it is a planning objective to bring about substantial changes in the distribution of income (i.e. if it is not presently regarded as optimal), then prevailing market prices lose their status as appropriate indicators of social valuations. There are substantial difficulties in interpreting this argument in the environmental context. It is not clear what redistribution is being adopted as a social objective nor, more importantly, is it at all self evident what change in valuation would be associated with any particular redistribution. Modifications of behavioural prices on this argument are more likely to reflect the arbitrary valuations of planners than the real valuation of individuals.

(ii) The second reservation arises where there are "*externalities of consumption*" and the individual's observed valuation does not properly represent the social valuation of the commodity. A good example of this is road safety. The available evidence suggests that individuals, in their observed behaviour, treat their own safety relatively lightly. They do not incur and therefore do not take into account the community costs associated with accidents or the grief in their own demise. Behavioural evaluations may therefore underestimate the social value of safety.

(iii) A not dissimilar argument is that of *pure paternalism*. This arises where the view is taken that the individual does not know what is good for himself. This may be a purely moral assertion or may reflect a genuinely superior knowledge of the probable consequences of an individual's action than the individual himself possesses. For instance, while individuals appear to dislike the smoke which vehicles sometimes produce (which is relatively harmless) they may be ignorant of, and hence indifferent to, more noxious fumes.

(iv) A fourth area is where there is "*jointness in consumption*". In some cases, e.g. defence, or police, the individual cannot sensibly purchase for his sole consumption any useful quality of the service in question. His valuation is non-marginal and depends on an assumption that everyone else will contribute to purchase for communal benefit the global amount. Behavioural evaluations in the market are thus impossible to obtain.

The conclusion that we are led to by these caveats is that there may be some aspects of the environmental problem which must essentially be politically administered and for which behavioural evaluation is not appropriate. But it is a corollary of the arguments from which this conclusion has been derived that the scope for an "administered price" approach is limited and susceptible to specific explanation. Moreover, even where the arguments are accepted it is still appropriate for the selection amongst alternative designs to be subject to tests of sensitivity to different administered evaluations.

This still leaves many aspects of environmental quality which impinge directly on individuals and for which behavioural evaluations appear to be appropriate. The problems presented here are essentially those of research technique. Probably the most common situation in which environmental valuations are taken into a direct market consideration is the determination of house prices. Unfortunately the house market is extremely complex and experience so far suggests that it is unlikely to yield any reliable and precise evaluation of specific environmental dis-benefits. More hopeful is the analysis of decisions on remedial treatment for specific dis-benefits such as sound-proofing though even here the trade off situation is far from simple.

In comparison with the trade off situations analysed in the evaluation of time savings, which are based on observation of simple and repetitive choice situations (e.g. choice of mode in journey to work), the real market environmental choice situations appear to be unmanageably isolated and complex. More recently, therefore, attempts have been made to simulate market choice situations in laboratory experimentation. The loss of realism is compensated by greater experimental control and the ability to generate data for analysis much more easily and cheaply. Early results of this approach have shown an encouraging consistency. Much more work remains to be done, however, in the development of research techniques in this difficult field.

CONCLUSION

In transport planning there does not yet appear to be any adequate bridge between general environmental objectives and detailed planning policies and project designs. It has been the argument of this paper that, because the problem is the inescapably economic one of husbanding resources, the quantification and evaluation of environmental effects is essential in building this bridge. Research along these lines is not yet well developed, and can never, in any case, completely eliminate the need for political judgements or design skills. But, so long as the status and limitations of behavioural evaluations in this peculiar area are borne in mind, an economic approach can contribute substantially to a more rapid and more effective incorporation of environmental objectives in design and policy formation.

ACKNOWLEDGEMENT

I would like to thank Miss Fleur Rees for her helpful comments and advice on this paper.

REFERENCES

1. C. H. Sharp and A. Jennings, "The costs and benefits of applying power to weight regulations to goods vehicles" *A Report to the Ministry of Transport* (1970).
2. A. R. Prest and R. Turvey, "Cost benefit analysis: a survey" *Econ. J.* (December, 1965).
3. N. Lichfield and associates, "Planning blight in social cost benefit analysis: the North Cross Route in Camden" *G.L.D.P. Inquiry Support Paper* S27/222 (1971).
4. Working Group on Research into Road Traffic Noise, "A review of road traffic noise" *R.R.L. Report LR* 357 (1970).
5. D. W. Robinson "The concept of noise pollution level" *Ministry of Technology N.P.L. Aero Report Ac* 38 (1968).
6. A. C. Whaffen and D. R. Leonard, "A survey of traffic induced vibrations" *R.R.L. Report L.R.* 418 (1971).
7. Road Research Laboratory, "Air pollution from Road Traffic Report" *R.R.L. Report LR* 352 (1970).
8. R. G. Hopkinson, "The quantitative assessment of visual intrusion" *J. Roy. Town Planning Inst.* (December, 1971).
9. N. E. Delaney *et al.*, "Propagation of traffic noise in typical urban situations" *N.P.L. Acoustics Report Ac* **54** (1971).
10. D. Crompton, "Traffic characteristics and environmental standards: problems of determining environmental capacity of streets". *Proc. P.T.R.C. Seminar on Environmental Standards* (May, 1970).
11. R. A. Waller, "Annoyance due to traffic noise and the effects of distance and traffic flows" *Proc. P.T.R.C. Seminar on Environmental Standards* (May, 1970).
12. Committee on the Third London Airport, Final Report. H.M.S.O., London (1970).

ON ASSESSING THE ENVIRONMENTAL IMPACT
OF URBAN ROAD TRAFFIC

FRANK EDWIN JOYCE and HUGH EDGAR WILLIAMS

Joint Unit for Research on the Urban Environment, University of Aston, Birmingham, U.K.

The objectives of urban transportation plans are increasingly being framed in environmental terms. This has resulted in part from the knowledge that the transport system can be manipulated to achieve desirable patterns of urban development and in the more negative sense from growing awareness of the disamenity that stems from urban road traffic. Discussion is restricted to the physical environmental impact of road traffic in terms of noise, pedestrian delay and hazard, visual intrusion, severance and atmospheric pollution. Recent research in the U.K. is examined. Special attention is drawn towards the planning problems that will result from increasing traffic on "secondary and local roads". The final section deals with problems presented by the lack of a comprehensive evaluative framework for seeking environmental goals and recommends the use of some interim measures.

THE CHANGING ROLE OF URBAN TRANSPORTATION POLICY

Examination of the objectives of urban transportation plans over the past few years reveals two distinct trends. Firstly, a movement away from objectives set in a purely highway planning context to consideration of the wider systems implications of the public and private transport mix; secondly, a growing awareness of the need to formulate and move towards broader and more complex environmental goals. The latter indicates concern both with the role that the urban transportation system plays in the achievement of acceptable patterns of urban development and with objectives directed to improving environmental quality. The impetus for these developments has resulted from a number of pressures technological, social and economic; not the least of which has been large scale public disquiet about the levels of traffic noise, atmospheric pollution, pedestrian hazard and delay that are present in large cities. This paper's concern is to review some of the recent research, principally in the U.K., which attempts to contribute to this aspect of policy. It focuses on the usefulness of the work to the urban planner and as such large areas of current, ongoing research in the physical and natural sciences which contribute to the wider ecological debate which are omitted.

To date public disquiet and professional and academic response has largely been directed at problems associated with motorways in urban areas. Work is at present in progress in a variety of organisations concerned with these issues under the guidance of the Urban Motorways Committee† set up by the U.K. Department of the Environment. These, however, are largely primary network problems: the environmental implications of traffic growth and distribution on roads lower in the hierarchy (secondary and local) have largely been ignored. It can be argued that the environmental impact will be more serious on the secondary road network.[2] It will certainly be more widespread for at present only 22·5 miles of urban motorway have been built in Britain (in the foreseeable future this could rise to over 200 miles) while in London alone there is at present 1000 miles of secondary roads straddled by shopping, commercial, industrial and residential land uses.

The planning and research issues at the primary level are principally remedial. The objective is to integrate the motorway within the urban fabric. Concern is focused on the "technological external diseconomies" (the disamenity factors) in terms of

† The Urban Motorways Committee was set up in 1969 by the U.K. Ministers of Transport and of Housing and Local Government with the following terms of reference:
  (i) to examine present policies used in fitting major roads into urban areas;
 (ii) to consider what changes would enable urban roads to be related better to their surroundings physically, visually, and socially;
(iii) to examine the consequences of such changes particularly from the points of view of:
    (a) limitations on resources, both public and private;
    (b) changes in statutory powers and administrative procedures;
    (c) any issues of public policy that the changes would raise;
(iv) to recommend what changes if any should be made.[1]

their scale and distribution and the need to develop acceptable design standards. The results of such research could form the basis for extending legal powers of land and property acquisition for environmental reinstatement or for some form of financial compensation. The problem turns upon the decision as to where the cut off point for any given "externality" lies and the difficulty of placing a value upon it. None of the work initiated by the Urban Motorways Committee has, as yet, been officially published. Roberts,[3] however, in a general review of the impact of urban motorways reports some cost comparisons of different forms of construction and remedial treatment for a route that passes through high density urban fabric; and also a technique currently being developed, which parallels cost effectiveness in character in so far as, for any given vertical alignment, impact can be predicted. Each type of impact will have a range of ameliorative possibilities ranging from insulation to redevelopment. If environmental standards are regarded as "given" then the method arrays the consequences in terms of the stated objectives.

The improvement of environmental quality, however, as an "objective function" comes into its own at the local road network level. This partly stems from the environmental area concept put forward by the Buchanan Report[4] in 1963 and partly from the argument that it is through the scale of provision of primary and secondary network roadspace that relief to local and residential roads can be achieved. An example of this approach may be seen in the Greater London Development Plan objectives which state that "the intention of the primary and secondary road network proposals of the G.L.D.P. is that they will permit the achievement of environmental benefits within the local networks".[5]

Praiseworthy as this kind of environmental objective may be it presents a range of difficulties to the urban planner. Primarily the problem is that we are not sure what environmental objectives we are trying to achieve. Some of the issues are reflected in the general debate on urban transportation investment appraisal. Gwilliam,[6] for example, while recognising the existence of disamenity factors, regards them as imponderable externalities. How much more formidable the problems become, then, when disamenity is to be treated not as an "externality" but as an "objective function". Measurement and prediction of the physical components of disamenity becomes a central issue together with the level of disamenity that people find unacceptable

and the value that those exposed to it would place upon its reduction. Other problems are implicit in the policy tools. For example, it is unlikely that the mere provision of increased capacity at the primary level would result in relief to other roads in view of the regenerative characteristics of urban traffic. Others still are technical such as the problem of forecasting traffic volumes at the micro scale where transportation modelling techniques may lack resolution.

This area, then, is highly sensitive in both policy and technical terms. Although the issues are often the subject of vigorous debate there is surprisingly little in the way of detailed research to show. Below we have attempted a review of progress and in so doing limit discussion to the physical dimensions of the problem.

SOME ENVIRONMENTAL PRESSURES

(i) *Traffic Noise*

The problem of noise is, perhaps, the impact of road traffic that has received most attention. Measurement is well defined in the physical sciences as is the use of the "A" weighting network, based on perceived loudness, and statistical analysis to condense the information. Many methods of manipulating such measurements have been developed in attempts to discover an index that relates the level of noise and its occurrences over time, with the degree of annoyance felt by those exposed to it. One of the earliest was the L_{10} noise level (the noise level exceeded for only 10% of a given period of time) which was used to suggest standards for noise reduction from traffic in urban areas.[7] Griffiths and Langdon[8] developed a "Traffic Noise Index" defined as

$$TNI = 4L_{A10} - 3L_{A90} - 30$$

where L_{A10} = the sound level in dB(A) exceeded for 10% of the time,

L_{A90} = the sound level in dB(A) exceeded for 90% of the time.

This was found to have a better correlation with annoyance than the L_{10} index.

Robinson[9] has reviewed this and many other indices that are used abroad and in the U.K. and has developed a single index that is capable of general application. He defines a "Noise Pollution Level" of the form:

$$L_{NP} = L_{eq} + 2 \cdot 56\sigma$$

where for most traffic conditions a good approximation is:

$$L_{eq} = L_{50} + \sigma^2/8 \cdot 68$$

where L_{50} = median noise level,
σ = standard deviation of noise level fluctuations over the period considered.

However, the prediction of the level and propogation of noise generated by traffic remains a difficult task. The degree of exposure depends upon traffic characteristics, (density, speed, composition and volume) distance from the noise source, the nature of the intervening surface, wind direction and the characteristics of the road. Thus any simple formula is not to be expected. However, Scholes and Sargent[10] have published a useful series of graphs for the calculation of noise exposure using a step by step procedure which makes successive corrections for traffic speed, distance from the road, etc. This is based in part upon the earlier theoretical and empirical work of Johnson and Saunders[11] whose measurements were carried out in the environs of roads with "free flow" conditions. The results are of limited applicability since free flow conditions are rare on heavily congested urban roads. Crompton and Gilbert,[12] however, have undertaken an extensive series of measurements of traffic noise in city streets and attempted to establish relationships between the noise level and the urban traffic conditions. The results of this work are as yet unpublished.

(ii) Pedestrian Delay and Hazard

As early as 1937 Adams[13] reported the results of both theoretical and empirical consideration of the relationships between pedestrian delay and traffic flow. The formula developed takes the form:

$$D = \frac{1}{N(e^{-NT})} - \frac{1}{N} - t$$

where D is the mean delay to pedestrians crossing the road (in seconds)
N is the traffic flow (in vehicles per second)
t is the minimum headway in a traffic stream that pedestrians will accept, in order to cross.

The formula assumes a random arrival pattern of traffic. Herein lies a difficulty for urban traffic flows on congested links of network with traffic signals, pedestrian crossings, etc., may not be random. Dowell[14] has examined the sensitivity of the formula to non-random conditions and has shown that delay to people accepting a 4-sec headway is about 80% higher than for a similar headway in random conditions for like volumes of traffic. This variation diminishes gradually until for a 10-sec gap acceptance the increase is only 10%. The percentage variation, however, may tend to overemphasize the absolute variation in seconds and Dowell, therefore, concludes that Adam's formula can generally be applied. It should be noted that the value of "t" is of considerable significance in the use of the formula particularly for high traffic flows. Pedestrians vary in gap acceptance and it is likely that in some situations pedestrians will not cross at all but be forced to undertake extensive diversions to reach a particular destination or accept gaps which involve high risks and a consequential social cost in terms of stress rather than delay. Adam's work indicated variations between 3·5 and 5 sec while Dowell's observations show that a representative value for minimum acceptable headway lies between 4 and 5 sec. However, for environmental evaluation he suggests the use of the 85th percentile value of 6·5 sec to allow for more vulnerable pedestrians.

Crompton,[15] in a study in Edinburgh, suggests that delay can best be expressed in terms of the percentage of pedestrians delayed (in an off-peak hour). A standard for office, shopping and residential streets of not more than 50% of the pedestrians wishing to cross the street at random times and locations being delayed was adopted. This was varied for special circumstances in individual streets and basically only applied where pedestrian demand was heavy. Environmental impact studies also require an assessment of the total numbers of pedestrians suffering delay or diversion. This at present largely involves analogue methods which merely hold constant observed pedestrian demand and distribution. Sandahl,[16] however, has attempted to simulate pedestrian demand in shopping areas by establishing its relationship with retail floor area and turnover, together with the centrality of the street in relation to the shopping area as a whole. Also Johnson[17] has applied a generation, distribution and assignment model on the transportation planning pattern to pedestrian movement in a small area of Birmingham.

Although other work has been carried out on pedestrian delay, principally by the Transport and Road Research Laboratory (T.R.R.L.),[18] this has been directed towards traffic engineering issues such as delay at traffic signals, pedestrian crossings etc., rather than at the issue of the environmental impact of traffic in the wider sense.

Clearly traffic flow involves danger to the pedestrian as well as delay but the calculation of the degree of risk involved is difficult because of the time period over which data must be collected to establish the relationship. Thus while data on the traffic parameters may be readily available it is often necessary to assume unchanging pedestrian demand.

Jacobs and Wilson[19] report that the overall risk to pedestrians crossing the road was found to increase with vehicle flow. A graph was produced relating vehicle flow to pedestrian risk and the following formula represents the regression line:

$$r = \frac{F}{10,000} - 0.03$$

where: r is the risk expressed as the number of pedestrian casualties in the 2·5 years divided by the pedestrian flow in 12 min (representative of the morning, evening and off-peak periods), F is the traffic flow in vehicles/hour surveyed for the same time periods.

Traffic speed is a further variable which influences pedestrian risk. The T.R.R.L.[18] have examined the effect of speed on gap acceptance by pedestrians and show that a slight increase in danger occurs up to 40 m.p.h. and a more significant increase above this figure. However, it is below 40 m.p.h. that most urban secondary and local roads will operate and the risk factors thus become difficult to evaluate. The speed variable may be handled either as an output from the traffic assignment process or input as a design standard.

Jacobs and Wilson's[19] work also indicates that the number of junctions in a network can have some effect on pedestrian safety as risk within 20 yd of a road junction appears to be just over twice that at other locations. There are, of course, other design factors which influence pedestrian safety but they are not directly relevant to environmental evaluation.

(iii) *Severance*

Severance is a complex issue which defies detailed analysis. It may take many forms, being simply the expression of pedestrian delay caused by high traffic volumes or the risk element which becomes so serious as to create psychological barriers between residential areas and local facilities. It is central to the problem of the competitive goals of accessibility and environmental quality. A measure of severance may be the degree to which cohesive communities are split or the reduction of catchment areas for small traders or indeed the extra journey time that would fall upon vehicle users that results from the definition of an hierarchical road structure.

(iv) *Visual Intrusion*

No generally agreed unit for the measurement of visual intrusion has yet been developed. Those suggestions that have been put forward, whether relating to the visual intrusion of vehicles themselves or of highway artefacts, are based on the concept of measuring the proportion of the individual's view that the intruding object occupies. Clearly a full analysis of screening effects should take into account the position of people in space and time, the position of the vehicles and the location of points and planes of interest to the onlooker. Such a task would be a formidable exercise and it is not surprising that more limited studies have been carried out. Eyles and Myatt[20] report a study that assumed the point of observation and the point of interest as fixed and then investigated the effects of varying elements connected with vehicles. Separate analysis was undertaken of the screening effects of four typical vehicles (car, van, lorry and bus) in varying locations in a given street. The levels of screening were calculated by three dimensional projective geometry. The calculation of the area of screening was carried out by defining two reference lengths at the object, one horizontal and one vertical. The angles which the lines subtend at the eye, and the relative positions of the observer, screening object and plane of interest, together determined the screened area in the plane of interest. Hopkinson's[21] work uses similar techniques although he has applied it to road structures rather than traffic. It also includes corrections for the nature of the human field of vision, giving reduced importance to static intrusions on the periphery of the field. He also introduces the "size constancy effect"† which causes anomalous results when this type of physical measurement is correlated with subjective response.

A large amount of development work is called for in this area. (Although some work by the Coventry Transportation Study Group[22] has been

† When an object that is known by experience to be very large is viewed from a distance the "size constancy effect" operates and it is subjectively assessed to be larger than would be expected from objective measurement of the angle it occupies in the visual range.

carried through to the prediction stage.) Two major difficulties are apparent. Firstly the field of view of the observer is not static, except perhaps in terms of a view from a window, and the quality as well as the quantity of intrusion is an important factor. The latter perhaps could be standardised either by reference to a group of professionals' evaluation of standard views, as in the work at Coventry[22] or that reported by Fines[23] at Sussex, or perhaps by evaluating individual preferences for visual attributes as in the work of Peterson and Neumann.[24,25]

(v) *Atmospheric Pollution*

The pollutants which result from road traffic include visible smoke, carbon monoxide, oxides of nitrogen, lead compounds and aldehydes.[26] Dockerty and Baley[27] comment that with the possible exception of some gases in the immediate environs of certain chemical works, no other air pollutant of the toxicity of carbon monoxide exists in such relatively high concentration in urban air. Fussel[28] has stated that 90% of carbon monoxide in the urban air is produced by motor vehicles. Despite this in the context of environmental evaluation of urban transportation projects there has been little theoretical or empirical work directed towards finding a relationship between CO levels and traffic characteristics. However, Bayley and Dockerty[29] have recently published the results of a small study which indicated carbon monoxide levels and traffic flows for a period of twenty four hours in one street in Birmingham. The rather more extensive studies undertaken by Crompton and Gilbert[12] have not yet resulted in any published relationships.

Sherwood and Bowers[26] observe that the general consensus of opinion in the U.K. is that air pollution from traffic is not a serious health hazard. This conclusion has been reached because no one has been able to demonstrate that, at the levels of concentration found in urban air, any of the pollutants have permanent harmful effects. There is, however, no dispute that traffic fumes are unpleasant, with minor health effects, and have an adverse effect on amenity in urban streets. It is, therefore, important to discover what value people place on the elimination of smoke and fumes in urban areas from traffic sources. This is obviously a difficult area of research and as such to date has largely been ignored.

(vi) *Loss of Daylight, Sunlight and Visual Privacy*

There has been little work on daylight, sunlight and loss of privacy related specifically to the effects of traffic proposals. Although the existing methods for calculating daylight and sunlight were developed for use in demonstrating the effect of one building upon another,[30] their use could be extended to cover the effects of highway structures and even traffic. Research work on visual privacy has also concentrated on relationships between buildings, specifically housing layouts. It seems doubtful that this could be directly adapted for the effects of traffic. Furthermore, existing regulations such as privacy distances are based, as Brierley and Ferguson[31] have pointed out, on very little data. It would appear that further studies of householders' responses to the invasion of visual privacy by traffic are required to form a basis for suggesting methods of measuring privacy.

SOME PROBLEMS OF EVALUATION

So far we have touched briefly upon some of the work that has attempted to quantify the scale and location of environmental problems. The usefulness, of these results, however, in the formation of policy depends upon the availability of a comprehensive evaluation framework. As Hutchinson[32] has pointed out, existing welfare theory-based methods fail to treat adequately the non-user dimensions of urban transportation investments. None of the conceptual frameworks available provide an acceptable operational mechanism at present, for resolving the conflicting objectives of accessibility and environmental quality. Despite the change in emphasis that planning policy seeks to achieve at the micro level from accessibility to amenity, the specification of amenity as an "objective function" eludes us.

The amenity issues are currently discussed as externalities of projects designed to reduce movement impedance. As such environmental evaluation, in the narrow sense, takes the form of impact analysis. The scale of the external costs may be regarded as reflected in a surrogate value determined by remedial costing. (The cost of environmental reinstatement in the urban corridor affected by motorway construction.) This may be a valid procedure where the rate of return to the system user and operator is high and remedial costs low but becomes suspect in situations where it is low and "intangible" environmental gains in other parts of urban area are claimed to make a project a viable investment. This then would argue for the definition of the environmental gains as the objective function. Otherwise we have no mechanism for determining whether the environmental gains we seek to achieve are worth the cost of achieving them. This, however, is clearly a technical utopia. Nonetheless, the problems of externality valuation and the work being carried out by a number of researchers may throw light on these issues. The avenue of development may lie in the search for surrogate values. Waller and Thomas,[33] for example, claim that individuals can and do put a limit on the price they are prepared to pay for environmental comforts such as central heating or double glazing. In practice, however, it is difficult to isolate the individuals response to the environmental variable under study. House price differential analysis has proved attractive to some workers but suffers from the nature of the house as a multi-product good making it difficult to isolate the effects of, for example, size and facilities, from location, the amenity of the area and social linkages.

Hoinville[34] has used gaming techniques for observing the way in which individuals "trade off" one environmental advantage against another. This work, however, parallels attitude surveys which as Foster and Mackie[35] have pointed out can be misleading as the translation of essentially qualitative results from a sample of people's attitudes to amenity issues to quantitative estimates of the value of changes in environmental amenity is an undoubtedly difficult step. Asking people how much they would be prepared to pay to avoid noise, for example, would be unreliable unless they have some idea of the amount of noise they seek to avoid. The recent efforts of the T.R.R.L. in setting up "REAL", a laboratory which attempts to simulate the physical effects of the road offers some potential for overcoming this problem.

CONCLUSION

Some research, then, is in progress on these issues but positive results are proving elusive. In the meantime the planner may usefully examine the approach pursued by C. Buchanan and Associates in Edinburgh (U.K.). Here the effect of alternative highway network proposals on the existing network was assessed using some of the techniques developed by Crompton and Gilbert.[12,15] Environmental capacities for given streets were defined based upon the effects that traffic has in environmental terms and the specific characteristics of the street. Such a procedure can allow either the comparison of alternatives as an impact study or the desired environmental standards expressed through restraint on capacity can be input as to the transportation modelling process as an initial constraint on possible "futures". Where specifically environmental objectives are sought project evaluation within the framework of a goals achievement matrix as advanced by Hill[36] may be helpful. It will stimulate the exposition of the real objectives of any project and make the distributional issues that are involved in the intuitive weighting of one project against another more explicit. Finally, it is interesting to speculate on the advantages that could be derived from the use of the approach pursued by Hoinville[34] in attempting to establish the relative "weight" that should be attached to any one objective by reference to community attitudes.

Some of the areas of research that we have reviewed may seem narrow and esoteric. They represent, however, early attempts to grapple with the issues of environmental quality. Urban highway provision is a sector of the economy which absorbs large resources for multi-purpose goals. We must be sure that we are achieving genuine returns for expenditure whether in terms of movement or environmental quality; at present we are not sure.

REFERENCES

1. Greater London Council, "Environmental effects of the construction of primary roads—illustrative examples" *G.L.D.P. Inquiry—Background Paper No. 383* (Oct., 1970).
2. Llewelyn-Davies, Weeks, Forestier-Walker & Bor and Ove Arup & Partners, *Motorways in the Urban Environment*. A British Road Federation Report, London (1971).
3. J. Roberts, "Impact" *Official Architecture & Planning* **35**(2), 85–91 (Feb., 1972).
4. The Steering Group and Working Group, *Traffic in Towns: a study of the long term problems of traffic in urban areas*. H.M.S.O., London (1963).
5. Greater London Council, "Proposals for secondary roads—illustrative examples" *G.L.D.P. Inquiry—Background Paper No. 443* (Feb., 1971).
6. K. M. Gwilliam, "The indirect effects of highway investment" *Regional Studies* **4**(2), 167–176 (Aug., 1970).
7. Sir A. Wilson, "Noise—final report of the committee on the problem of noise". *Cmnd 2056*, H.M.S.O. London, (1963).
8. I. D. Griffiths and F. J. Langdon, "Subjective responses to road traffic noise" *J. of Sound and Vibration* **8**(1), 16–32 (July 1968).

9. D. W. Robinson, "An outline guide to criteria for the limitation of urban noise" *Aeronautical Research Council Current Papers CP No. 112.* H.M.S.O., London (1970).

10. W. E. Scholes and J. W. Sargent, "Designing against noise from road traffic" *Building Research Station Current Paper 20/71* (May 1971).

11. D. R. Johnson and E. G. Saunders, "The evaluation of noise from freely flowing road traffic" *J. of Sound and Vibration* 7(2), 287–309 (March 1968).

12. D. H. Crompton and D. Gilbert, "Traffic and the environment" *Traff Engng Control* 12(6), 323–326 (October, 1970).

13. W. F. Adams, "Road traffic considered as a random series" *J. Inst. Civ. Eng.* 4, Paper 5073, 121–130 (1936–37).

14. T. A. R. Dowell, "Some studies of pedestrian delay and noise in relation to traffic flow" *J. Inst. Munic. Eng.* 94, 264–268 (Aug., 1967).

15. D. H. Crompton, *Edinburgh, Environmental Capacities of Central Area Streets.* Transport Section, Department of Civil Engineering, Imperial College, London (July, 1971).

16. J. Sandahl, "A pedestrian traffic forecast model for town centres" Paper given to *PTRC Seminar: Urban Traffic Research*, Feb. 29–March 3. Planning & Transport Research Computation Limited, London (1972).

17. K. Johnson, "A pedestrian model based on a home interview survey" Paper given to *PTRC Seminar: Urban Traffic Research*, Feb. 29–March 3. Planning & Transport Research Computation Limited, London (1972).

18. Road Research Laboratory, *Research on Road Traffic.* H.M.S.O., London (1965).

19. G. D. Jacobs and D. G. Wilson, "A study of pedestrian risk in crossing busy roads in four towns" *Road Research Laboratory Report LR 106* (1967).

20. D. Eyles and P. Myatt, "Road traffic and urban environment in Inner London (a study of LTS Zone 277)" *Research Memorandum R.M. 250.* Department of Planning & Transportation, Greater London Council (Sept., 1970).

21. R. G. Hopkinson, "The quantitative assessment of visual intrusion" *J. Town Planning Institute* 57(10). 445–449 (Dec., 1971).

22. D. G. Leyland and D. K. Foster, "Visual intrusion of urban motorways—a quantitative assessment" *Roads and the Environment—Project Reports 1971.* Coventry Transportation Study (1971).

23. K. D. Fines, "Landscape evaluation: a research project in east Sussex" *Regional Studies* 2(1), 41–55 (Sept., 1968).

24. G. L. Peterson "A model of preference: quantitative analysis of the perception of the visual appearance of residential neighbourhoods" *J. Reg. Sci.* 7(1), 19–31 (1967).

25. G. L. Peterson and E. S. Neumann, "Modelling and predicting human responses to the visual recreation environment" *J. Leisure Res.* 219–237 (1968).

26. P. T. Sherwood and P. H. Bowers, "Air pollution from road traffic—a review of the present position" *Road Research Laboratory Report L.R. 352* (1970).

27. A. Dockerty and E. Bayley, "Carbon monoxide pollution of urban air" *Traff. Engng. Control* 11(12), 594–597 (April, 1970).

28. D. R. Fussell, "Atmospheric pollution from petrol and diesel engined vehicles" *Petroleum Rev.* 192–202 (July, 1970).

29. E. Bayley and A. Dockerty, "Traffic pollution of urban environments" *Roy. Soc. Health J.* 92(1), 6–11 (Jan./Feb. 1972).

30. Department of the Environment, "Planning for sunlight and daylight" *Planning Bulletin No. 5.* H.M.S.O., London (1964).

31. E. S. Brierley and J. F. Ferguson, "Choice in housing design" *Building* 220(6683), 25/127–25/128 (June, 1971).

32. B. G. Hutchinson, "Structuring urban transportation planning decisions: available social science constructs" *Environment and Planning* 2, 251–265 (1970).

33. R. Waller and R. Thomas, "The cash value of the environment" *ARENA* 82(904/9), 164–166 (1966).

34. G. Hoinville, "Evaluating community preferences" *Environment and Planning* 3, 33–50 (1971).

35. C. D. Foster and D. T. MacKie, "Noise: economic aspects of choice" *Urban Studies* 7(2), 123–135 (June 1970).

36. M. Hill, "A goals achievement matrix for evaluating alternative plans" *J. Amer. Inst. Planners* 34(1), 19–29 (Jan. 1968).

PEOPLE, TRANSPORT SYSTEMS, AND THE URBAN SCENE: AN OVERVIEW—I

C. A. O'FLAHERTY

Institute for Transport Studies, University of Leeds, Leeds LS2 9JT, U.K.

This paper—which is divided into two parts—examines the problem of transport in towns, with particular relevance to the relationship between the motor vehicle and the urban environment.

This, the first part, briefly traces the development of urban transport to the situation where it is today. In so doing particular attention is paid to the motor car, and to its effects on the environment, after which discussion is turned toward how best to reconcile the quality of life in urban areas with long-term land use and transport developments. Future developments in relation to transport in towns are discussed at length, and it is concluded that no radical developments in movement technology are likely within the foreseeable future.

INTRODUCTION

It is sometimes said that our urban travel problems are not really new, that historically they have always been with us. To a certain extent, I suppose that this is true. Certainly, when we read history, we come across many references which might suggest this. For example, King Sennacherib of Assyria, who ruled about 70 B.C., is reputed to have forbidden parking on the main road of his capital city and enforced this decree by directing that any illegal parker on the 78-ft wide royal road should be put to death by being impaled upon a pole in front of his home. (Fortunately, our parking regulations are less rigidly enforced in this day!)

As a general rule, however, urban travel was not *considered* to be a problem until the advent of the Industrial Revolution. Prior to then travel within towns was, on the whole, by horse or foot—and towns were generally restricted in size to a (maximum) radius of about three-quarters of an hour's travel time from their centres. Came the Industrial Revolution, and the situation radically changed—indeed it might be said that civic life and form then changed almost beyond recognition. A great wave of migration from the country to the town commenced—and this was accompanied by the beginning of what we now know as the "population explosion" (see Figure 1).

This past history is of importance in that we should recognize that our towns and cities of today were never planned—but rather that they are the legacies of the relatively sudden growth in urban structure which then took place.

As urban populations grew, distances and times involved in daily travel into and within urban areas

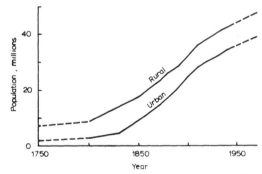

FIGURE 1   Urban and rural populations of England and Wales (based on data in Ref. 1).

likewise increased—and the need for flexible public transport facilities began to be recognized:

—July 4, 1829 saw the initiation of mass public transport in towns when Mr. George Shillibeer, the "father" of the British public transport system, started his horse "omnibus" service in London between the village of Paddington and the Bank. (This particular service was not very successful, and Shillibeer ended his career as an undertaker. This, the first omnibus in Britain, was able to carry 20 passengers; it picked up and set down its passengers in the street—a procedure that did not, in fact, become legal until 1832.)

—April 22, 1833 saw Mr. Walter Hancock testing

his steam omnibus (capable of carrying 14 passengers) with the intention also of starting a regular service between Paddington and the Bank.

—August 30, 1860 saw the beginning of the end of the horse-bus when Mr. George Train initiated his horse-tram service in Birkenhead. The horse-tram ran on a rail set in the ground, and so it could carry more people more comfortably and more quickly than the horse-bus.

—January 10, 1863 saw the opening of the world's first underground railway, the first part of the Metropolitan (steam) Railway which ran between Paddington and Farringdon.

—September 29, 1885 saw the opening of the first electric street tramway at Blackpool—and so initiated the death of the horse-tram. (Indeed, it can be also said that the electric tram had a really tremendous effect on town development, for within 25 years nearly every major town in Britain had its own network of electric tramways. Typically, these tramways radiated outward from the central area and serviced settlements clustered along these routes. Thus transport development began to be clearly associated with now familiar pattern in which residential densities decline from zone to zone from the centre outwards.)

—April 12, 1903 saw the start of the first municipally-operated motor bus service at Eastbourne. (A major feature of the motor bus was that it brought the suburbs within easy reach of the town centre. Furthermore, unlike the tram, it was not forced to stay on fixed routes, and so it could service housing not astride the radial arteries.)

The advent of the motor car about the end of the 19th century had little significant effect on mass movement or town development for some time for the very practical reason that in those early days it could only be enjoyed by the wealthy. (Of much more importance to the general populace of that time was the "low" bicycle.) It was not until after World War 1, when the mass production methods developed in America were applied in Britain, that the price of the motor car was sufficiently reduced for it to start to become the ordinary man's carriage. It is then that the era of "personalized travel" began in Great Britain.

THE MOTOR CAR AND ITS EFFECTS

Henry Ford once said in the United States:

"I will build a motor car for the great multitude . . . so low in price that no man . . . will be unable to own one—and enjoy with his family the blessing of hours of pleasure in God's great open spaces. . .."

Nearly 13 years ago, the U.S.S.R. Premier, Nikita Khrushchev visited the United States and was taken to see one of the largest motor car assembly plants. While, there, Mr. Khrushchev's hosts made a point of showing him the acres and acres of car parks which were covered with the personal cars of the factory workers. His comment: "What a waste".

Who was right—Mr. Khrushchev or Mr. Ford? Or was neither correct?

Personal Mobility

There is no doubt but that the motor car is the most marvellous instrument of personal mobility known to man. It is, in a sense, his own private "magic carpet" in that it allows him to go where he likes, when he likes. It picks him up at his doorstep and deposits him at the doorstep of his destination. The man with a car need not look up a timetable or wait in the rain at a bus stop or stations for public transport. He can see parts of his and other countries which he never could see before. His children can be brought to school or to the dentist in comfort. His wife need not shop every day, for now larger shopping loads can be carried home at any given time. Above all, perhaps, is the fact that the private car satisfies his need to make individual decisions regarding transport in his own time, and at his own convenience.

These are but a very few of the advantages which the private car provides its owner. Whether one likes the car or not, they must be admitted. Certainly the ordinary man seems to think so—and he has expressed his opinion in terms of hard cash. In 1949, there were 2·13 million private cars on Britain's roads;[2] in 1959, the number was 4·97 million, and in 1969 this had more than doubled again to 11·23 million.

The mobility provided by the motor car is well illustrated in Table I; it shows that the effect of increasing car ownership is to increase travel from each household. This table also provides the very interesting information that the increase in travel is mostly for *non-work* purposes, which implies that these trips were made only because of the availability of the car.

Accidents

There are unfortunately, a number of clearly undesirable byproducts of the development of the motor vehicle. Not least amongst these is the great number of accidents involving this "magic carpet".

TABLE I

Trip generation, by mode and purpose, per household per day for households in London with one employed resident per household in an urban area of average rail and bus accessibility[3]

Purpose	Household income (£)	Private transport			Public transport			All modes		
		0-car	1-car	Multi-car	0-car	1-car	Multi-car	0-car	1-car	Multi-car
Work	0–1000	0·2	1·2	—	0·9	0·4	—	1·1	1·6	—
	1000–2000	0·3	1·2	1·6	1·1	0·6	0·5	1·4	1·8	2·1
	>2000	—	1·0	1·5	—	0·5	0·2	—	1·5	1·7
Non-work	0–1000	0·2	2·0	—	0·8	0·4	—	1·0	2·4	—
	1000–2000	0·3	2·7	5·7	1·4	0·7	0·3	1·7	3·4	5·5
	>2000	—	3·9	8·4	—	1·1	0·9	—	5·0	9·3

Figure 2 expresses in graphical form the annual road casualty figures for this country since 1928.

Looking at this figure, one is tempted to "point with pride" at the manner in which different casualty rates have been reduced since World War 2, although the number of vehicles on the roads has more than quadrupled over the same period of time. There is no doubt but that much of the improvement in the accident statistics may generally be associated with improvements in highway and traffic engineering, e.g. better road and junction design, better road marking and signing, better road lighting, better testing of vehicles, better vehicle design including the fitting of seatbelts, better public education, and better traffic law including that relating to drinking and driving.

Then I look again at this figure and I realize that the statistics portrayed there represent human lives and human errors. The magnitude of these errors is perhaps best summed up in Table II which shows the total numbers of the two main types of casualties resulting from road accidents at different intervals since World War 2. And there is one thing we can be sure about—these numbers will be increased upon every year into the foreseeable future.

It is indeed a horrifying thought that, on the basis of the figures for the past 20 years, the next 20 years will see, *at the very least*, 1·5 million people killed or seriously injured on the roads of this island, and a total of at least 6·5–7 million injured. Furthermore, on the basis of current experience, it is likely that some 65% of those killed and seriously injured (and about three-quarters of all casualties) will be involved in accidents in built-up areas—and a very high proportion of these will involve pedestrians.

Noise

The mechanical sources of noise generally encountered in modern towns are road and rail traffic, aircraft and industrial noise. Of these the most important is undoubtedly road noise. Representative existing external noise climates associated with road traffic are tabulated in Table III.

In 1963, the Wilson Committee[5] carried out a most thorough examination of the noise problem created by motor vehicles. In so doing it concluded that the internal noise level by day in buildings in towns should not exceed 50 dB; this means, in fact, that the external noise level should not exceed 70–80 dB (to allow for the insulating effect of a building with closed windows). In other words a maximum noise level of 75 dB(A) at a distance of

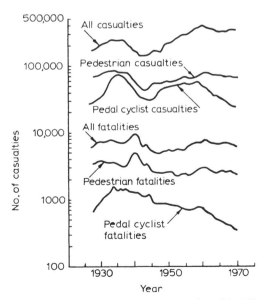

FIGURE 2   Road casualties in Great Britain, 1928–1969.

TABLE II
Road casualties in Great Britain, 1949–1969[4]

| Year | 1949 | | 1959 | | 1969 | |
Casualty	Killed	Injured	Killed	Injured	Killed	Injured
Pedestrians	2315	50,767	2520	65,160	2956	80,119
Pedal cyclists	842	42,574	738	51,333	411	23,950
Other persons	1616	78,665	3262	210,440	4016	241,742
All	4773	172,006	6520	326,933	7383	345,811

about 20 ft from a building whose windows are shut, is the most that should be tolerated. If the windows are open, then the maximum value to be tolerated is about 70 dB(A).

Despite the very many and varied investigations which have been carried out on this subject, relatively little is known about the real effect of road traffic noise upon the individual person—and much less about its effect upon the community. Certainly, there is no evidence to suggest that illness, either physical or mental, is caused by road traffic noise.[6] Thus, if one is to judge its effect by observed community action, complaints or increases in illnesses, then it would appear that road traffic noise is not a great nuisance. However, if one is to judge it by noting difficulties in holding conversations, or in listening to the radio or television, or in getting to sleep, etc., then traffic noise is certainly a major intruder upon the quality of life in built-up areas.

It is likely that the road traffic noise problem in towns as a whole will get significantly worse before beginning to get better—at least it would appear that several factors are now combining to worsen the situation, e.g. an ever-increasing number of vehicles with greater horsepower used on the road, the improving of old (or the building of new) main roadways which bring high speed heavy traffic into residential and commercial areas, the construction of elevated roads (which worsen the problem, e.g. the Western Avenue extension connecting central London with the A.40), and the construction of high buildings (which are particularly susceptible).

Air pollution

Today there are no streets carrying appreciable amounts of traffic that are really free from odours and fumes. This is true for many residential streets, as well as for roads in busy central areas of towns.

TABLE III
Existing external noise climates

Group	Location	Noise climate [dB(A)]†	
		Day 8 a.m.–6 p.m.	Night 1 a.m.–6 a.m.
A	Arterial roads with many heavy vehicles and buses (kerbside)	80–68	75–50
B	1. Major roads with heavy traffic and buses	75–63	61–49
	2. Side roads within 45–60 ft of roads in Groups A or B1 above	75–63	61–49
C	1. Main residential roads	70–60	55–44
	2. Side roads within 60–150 ft of heavy traffic routes	70–60	55–44
	3. Courtyards of blocks of flats, screened from direct view of heavy traffic	70–60	55–44
D	Residential roads with local traffic only	65–56	53–45

† Noise climate is the range of noise level recorded for 80% of the time. The level exceeds the higher figure for 10% of the time and is less than the lower figure for 10% of the time.

TABLE IV
Typical concentrations of pollutants in vehicle exhaust gases[7]

Engine operation	Idling		Accelerating		Cruising		Decelerating	
Type of engine	Diesel	Petrol	Diesel	Petrol	Diesel	Petrol	Diesel	Petrol
Carbon monoxide (%)	Trace	7·0	0·10	3·0	Trace	4·0	Trace	3·0
Hydrocarbons (ppm)	220	820	110	700	55	500	160	4400
Oxides of nitrogen (ppm)	60	30	850	1050	250	650	30	20
Aldehydes (ppm)	10	30	20	20	10	10	30	300
Sulphur dioxide (%)	—	—	—	—	—	—	—	—

And, as with noise, because it has happened gradually, air pollution is tolerated and accepted.

Internal combustion engines burn fuels that are compounds of carbon and hydrogen. Ideally, the fuel should be burned completely in air to give only carbon dioxide and water—and there would then be no air pollution problem from traffic. This ideal cannot be obtained with motor vehicles, and so pollutants are produced; they include carbon monoxide, smoke, hydrocarbons, oxides of nitrogen, oxides of sulphur, lead and aldehydes. Typical average concentrations of the different pollutants in the exhaust gases of vehicles are shown in Table IV.

As far as is known at this time, the most dangerous emission from a motor vehicle is *carbon monoxide*. Carbon monoxide is toxic; it combines readily with the haemoglobin in the blood and, if in sufficient concentration, can cause death. Early symptoms of danger are headache, nausea and giddiness. It is reported[7] that roadside measurements made in London have shown carbon monoxide concentrations of 200 ppm for short periods, and average values of 50 ppm over a 10-min period. On one occasion 360 ppm was found on the pavement of Oxford Circus. These figures may be compared with the maximum allowable concentration of 100 ppm for an exposure period of 8 h in workplaces in this country. Concentrations of well over 100 ppm have been measured inside cars. The situation is most serious when the engines of stationary or slow-moving cars are idling, e.g. in congested town streets where there is little air movement. In such situations, the carbon monoxide can reduce visual acuity, cause sleepiness and lack of concentration, and may be a contributory cause to accidents—both by motorists and pedestrians.

Unburnt hydrocarbons get through all types of engines and are also emitted through the exhaust. They are significant in that they (a) take part in the photo-chemical reaction which gives rise to smog of the Los Angeles type, and (b) they may induce headaches and drowsiness. The Los Angeles smog is produced only in conditions of bright sunshine, and hence we do not have to worry too much about it in Great Britain. The fact that some hydrocarbons have been shown to cause cancer in animals has led to the allegation that they may contribute to the increases in lung cancer; at present, however, there is no conclusive evidence on which to blame either diesel or petrol engined vehicles for the increase in deaths from lung cancer.

The maximum concentrations of *oxides of nitrogen* found in London streets—at 0·05–1·5 ppm —is far below that which is considered to be the maximum safe level in industry, and they are therefore not thought to be medically significant at this time. High concentrations, i.e. 5–15 ppm, of nitrogen dioxide can cause permanent damage to the lungs—however, at this time it is considered that, in general, there is no direct medical hazard from oxides of nitrogen from motor vehicles.

Lead compounds added to petrol are also emitted in the form of halides. As yet, however—and although lead is a serious poison—there is no firm evidence to indicate whether the very tiny amounts emitted by motor vehicles have any long term effects on people's health.

The *aldehyde compounds* are not considered to be medically harmful either. At this time, their main fault is that they account for the characteristic smell of petrol and diesel exhaust fumes.

In summary, it can be said that from a health aspect carbon monoxide from petrol vehicles is a definite possible health hazard. The black smoke from diesel engines is also undesirable—not because of a health hazard but because it is unpleasant and is a road accident risk, particularly in traffic congestion or in narrow or enclosed roads where it may obscure driver view. There then remains, of course, the possibility (on which medical work is proceeding) that some pollutants may interact in the human body to produce detrimental effects which are as yet unknown.

Visual intrusion

An environmental 'dis-amenity' which, unfortunately, receives very little attention is the aesthetic deterioration associated with vehicle intrusion. Perhaps the best way to describe this particular effect is to quote directly from the Buchanan report[8] regarding the visual consequences associated with the motor car in towns:

". . . the crowding out of every available square yard of space with vehicles, either moving or stationary, so that buildings seem to rise from a plinth of cars; the destruction of architectural and historical scenes; the intrusion into parks and squares; the garaging, servicing and maintenance of cars in residential streets which creates hazards for children, trapping the garbage and the litter and greatly hindering snow clearance; and the indirect effect of oilstains which render dark black the only suitable colour for surfaces, and which quickly foul all the odd corners and minor spaces round new buildings as motor cycles and scooters take possession. There is the other kind of visual effect resulting from the equipment and works associated with the use of motor vehicles: the clutter of signs, signals, bollards, railings, and the rest of the paraphernalia which are deemed necessary to help traffic flow; the dreary, formless car parks, often absorbing large areas of towns, whose construction has involved the sacrifice of the closely knit development which has contributed so much to the character of the inner areas of our towns; the severing effects of heavy traffic flows; and the modern highway works whose great widths are violently out of scale with the more modest dimensions of the towns through which they pass."

These words received very great attention when they were first said in 1963. Today, the words are still said, but unfortunately very many of the actions which they suggest are necessary have yet to be implemented.

Accessibility

Accessibility is a reflection of the ease with which a person can travel from an origin to a destination. One simple way of determining the accessibility of a destination via different transport modes is to compare their door to door travel times at the time of day under consideration.

Smeed[9] has shown (see Figure 3), for example, how buses *and* cars using the same roadways in London can have their average journey times increased as a result of increasing percentages of

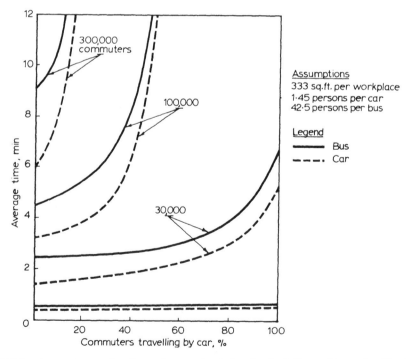

FIGURE 3   Average journey time for road commuters to London (excluding walking and waiting times,[9]

TABLE V
Passenger transport in Great Britain, 1959–1969[2]

Transport mode	1959		1961		1963		1965		1967		1969	
	Pass. miles $\times 10^9$	%	Pass. miles $\times 10^9$	%	Pass. miles $\times 10^9$	%	Pass. miles $\times 10^9$	%	Pass. miles $\times 10^9$	%	Pass. miles $\times 10^9$	%
Rail	25·5	17	24·1	14	22·4	12	21·8	11	21·1	9	21·6	9
Road												
Public service vehicle	44·1	29	43·1	26	41·5	23	39·2	19	37·0	16	35·7	15
Private transport	82·1	54	99·8	60	115·5	65	144·7	70	167·9	75	184·0	76
All	151·7	100	167·0	100	179·4	100	205·7	100	226·1	100	241·1	100

Note also that all road vehicles travelled a total of 122×10^9 vehicle-miles, including 94×10^9 car-miles, in 1969: this was almost equally divided between urban and rural roads. Trunk and principal roads, which comprise about 14% of the road mileage, accounted for about 70% of the total vehicle mileage.

people using their cars for work trips. No one who has sat in a car or a bus in ever-getting-longer traffic jams over the past decade will doubt the validity of the work reported in this diagram.

There is ample evidence to the effect that, given a free choice, most people will use their motor cars in preference to any form of public transport. This, for example, is reflected in the passenger mileage statistics shown in Table V. While these data reflect only national trends with regard to passenger movements, there is ample evidence elsewhere to show that there has been also a clear and continued swing away from the use of public transport in towns; the only points to note here are that the

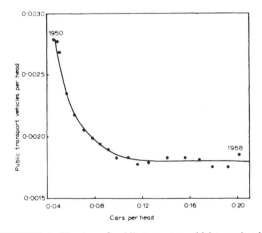

FIGURE 4 Number of public transport vehicles per head and cars per head in Great Britain, 1950–1968.[11]
Note: Public transport as used in this figure includes taxis and private hire cars, as well as public service vehicles.

amount of the decline has varied from town to town, and that the peak year for public transport passengers (prior to the decline) was much later in some towns than in others (see, for example, data in Ref. 10). At the present time it would appear that this swing with respect to passenger mileage and passenger journeys is likely to generally continue, albeit at a slow rate. On the other hand, the number of public transport vehicles per head of population may well have reached its base level (see Figure 4), and may not go much lower.

The concept of accessibility may also be extended here to include the ability to park swiftly and safely. This is quite important at shopping locations where the availability of parking facilities determines to a large extent the viability of retail activity. And there can be little doubt but that there is great difficulty in obtaining parking even for a short period of time in the central areas of very many towns today.

Land use

That the motor vehicle has had, and will continue to have, a major effect upon land usage within urban areas is very well recognized.

For example, there is the flight of home-dwellers to the suburbs, a product probably mainly of slum clearance and higher incomes, but at least partially made possible by the mobility given by the motor vehicle. There is also the movement of industry to sites adjacent to good road facilities. (This latter movement has been generally encouraged by town planning practice which, in the past, has generally favoured the development of the centripetal type of town—and indeed city region—of the idealized form shown in Figure 5.)

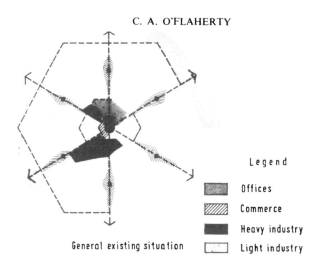

Legend

Offices

Commerce

Heavy industry

General existing situation   Light industry

● City centre
■ District centres
○ Neighborhood centres
---- Primary motor roads
- - - Secondary motor roads
—— Public transport route only

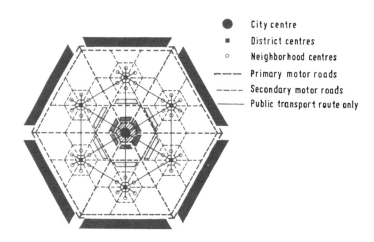

Idealised centripetal town

FIGURE 5   General town planning practice with respect to land use and the transport routes.

There is also the general decrease in the number of jobs in the central areas of the big towns, e.g. see Table VI. Again, while it cannot be said just how much of this is a reflection of transport problems, there is no doubt but that there is a genuine fear amongst central area employers that unless traffic congestion is reduced, the trend will become even more definite.

There is the general trend and continuing pressure for the establishment of shopping centres in suburbia. Of the factors listed in Table VII as influencing this trend, there is little doubt but that the total travel time from the home is the main

TABLE VI
Changes in employment in five central business districts, 1961–1966 (reported in Ref. 9)

Central business district	Decrease in employment	
	1000's	%
London	89	6
Birmingham	11	8
Liverpool	16	10
Manchester	25	15
Newcastle-upon-Tyne	Not significant	

TABLE VII

Factors influencing the decision whether to shop in the central area or in an outlying shopping facility

	Central area	Suburban centre
For	1. Greater choice of competitive goods 2. Opportunity of doing several different types of errands on one trip 3. Ease of accessibility by public transport 4. Lower prices	1. Closer to home 2. Easier parking 3. More convenient shopping hours
Against	1. Difficulty in parking 2. Very crowded 3. Traffic congestion	1. Fewer types of business 2. More limited selection of competitive goods 3. Poor accessibility by public transport 4. Prices are higher

advantage of the suburban centre. This again illustrates why city centre merchants are concerned with improving the Central Area-oriented transport facilities and, particularly, the parking facilities for shoppers in Central Areas.

FUTURE TRENDS: NUMBERS OF PEOPLE AND VEHICLES

People

Obviously any consideration of the future relative to passenger transport in towns must take into account any growth in the numbers of people. While predictions of the future population of Great Britain are fairly notorious for their variations, they all have in common that the population is going to increase. The most current predictions are given in Figure 6 (see 1970 data). While it is clear from this figure that there is no "true" answer as to what the population will amount to in any given future year, it is equally clear that the present number is going to be increased upon quite significantly—provided, of course, that there are no national catastrophies in the meantime.

Figure 7 shows that by far the greater part of this increased population will belong to the car-owning/car-driving ages.

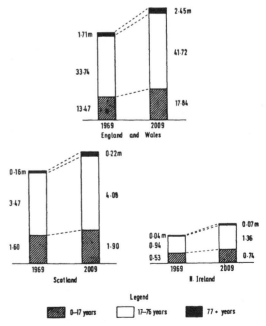

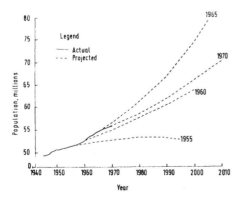

FIGURE 6 Population growth in Great Britain and Northern Ireland, 1945-2010.[13]

FIGURE 7 Population of Great Britain and Northern Ireland: 1969-2009 compared (based on data in Ref. 13).

TABLE VIII
Distribution of the population in the urban areas of England and Wales, 1969
(based on data in Ref. 12)

Population group × 1000	No. of urban areas		Population		
	Individual	Cumulative	Aggregate, × 1000	% of total	Cumulative (%)
Over 5000	1†	1	7703	15·8	15·8
4000–5000	—	1	—	—	15·8
3000–4000	—	1	—	—	15·8
2000–3000	2†	3	4874	10·0	25·8
1000–2000	2†	5	3069	6·2	32·0
750–1000	1†	6	840	1·7	33·7
500–750	1	7	529	1·1	34·8
250–500	8	15			
150–250	9	24	6001	12·3	47·1
100–150	18	42			
75–100	26	68			
50–75	49	117	4795	9·8	56·9
25–50	134	251			
Less than 25	504	755	10,579	21·7	78·6‡

† Conurbations.
‡ Remaining 21·4% reside in Rural Districts.

Where will these extra millions of people live? This can probably be assessed by taking as an example the current population distribution for England and Wales for the year 1969; this distribution is given in Table VIII.

The first point to be noted about Table VIII is that it emphasizes that Britain is essentially an urban nation, and that the great majority of people live in urban areas. (In fact, if anything, Table VIII under-estimates the proportion living under urban conditions since the outskirts of many urban areas lie within Rural Districts.) Furthermore, approximately one-third of the population of England and Wales lives in the five conurbations (see also Table IX).

In theory, all of the increases with regard to population in future years could be fairly easily catered for if the extra numbers could be distributed throughout all or a great number of the urban areas. In practice, however, this is not what will take place since people cannot and should not be told where to go and what to do; what will likely happen instead is that the bigger urban areas, particularly the conurbations, will take much more than their proportionate shares of these extra numbers.

TABLE IX
Basic characteristics of the conurbations in Great Britain

Conurbation	Registrar General's definition			Transportation study definition			
	Area (mile²)	Population in 1967	Density (persons/mile²)	Survey (year)	Area (mile²)	Population	Density (persons/mile²)
London	616	7,880,760	12,793	1962	941	8,827,000	9380
W. Midlands	269	2,446,400	9094	1964	376	2,529,000	6726
SELNEC	380	2,451,660	6452	1965	413	2,595,801	6285
Merseyside	150	1,368,630	9124	1966	158	1,453,333	9198
Tyneside	90	849,370	9437	1966	322	1,432,400	4448
Clydeside	300	1,764,430	5881	1964	454	1,929,000	4249
W. Yorkshire	485	1,730,210	3567	1966	1000	2,098,770	2099

It is useful here to briefly comment on conurbations for, as is noted later, it is within these that the very major planning and transport and environmental problems arise. At the present time there are six officially recognized conurbations in Great Britain (the sixth is the Central Clydeside conurbation in Scotland); it is likely, however, that this number will be added to in the future. These conurbations have been described[15] as consisting

"... of groups of urban settlements more or less in a state of coalescence which have developed over a period of time, usually but not always around a central dominant city."

London, in particular, has a structure which is strongly dominated by the central city, and this is emphasized by the form of its communication system which tends to be strongly radial and focussing on the centre. In the case of the West Midlands, South East Lancashire, Merseyside and Tyneside, the central city theme is less dominant, while the West Yorkshire conurbation consists essentially of a conglomeration of free-standing towns and cities. In all of the conurbations, the central cities are tending to lose population while the regions containing them are gaining. (Since 1961, for example, the Greater London conurbation has grown at the rate of almost 100,000 people every year.)

Motor cars

If it is accepted

(1) that people's standards of living will continue to rise,
(2) that people now regard "personal mobility" as an integral part of a high standard of living,
(3) that the numbers of people on this island will continue to grow, and
(4) that it is unlikely that any new form of personal transport will supplant the motor car within the forseeable future,

then it can reasonably be expected that the numbers of motor vehicles are likely to increase significantly in future years. An indication of the extent to which these numbers could grow is revealed in the projections made in 1969 by the Road Research Laboratory; these are shown graphically in Figure 8. Two of the more important assumptions utilized in preparing this diagram are (a) that Britain's population will be 71·7 million in the year 2010 (the Government Actuary now proposes a figure of 68·7 million) and (b) that the parameters defining the elements of vehicle ownership and use, etc. during future years will be quantitatively in accordance with those applicable now.

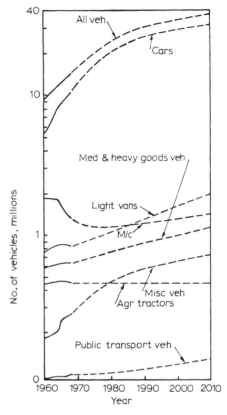

FIGURE 8 Actual numbers of vehicles in Great Britain 1960–1968, and forecasts 1969–2010.[11]

From what has been said previously with regard to the population distribution, it is clearly of interest to have an appreciation of the manner in which these large vehicle numbers, and particularly the numbers of cars will be distributed throughout the land. In 1968, there were 0·20 cars per head of population in Great Britain; by 1985 it is anticipated that this ratio will have become 0·39, while by 2010 it is expected to be 0·45 (at saturation level). These are the national projections; the extreme possibilities suggested by the Road Research Laboratory for various "environmental" locations in Great Britain for the year 2010 are as follows:

Rural counties	0·50 cars per head
Typical counties	0·45–0·50
Medium-sized towns	0·40–0·45
Conurbations	0·35–0·40
Inner London	0·30–0·35

Note in this respect that the large urban areas have relatively lower predicted car ownership values, thus reflecting a future heavier reliance on public transport in those areas.

STRATEGY FOR THE LONG-TERM FUTURE

It is probably true to say that the urban transport problem will never be truly "solved"—simply because life being what it is, one form of a problem is generally succeeded by another. Nevertheless, one must proceed on the hypothesis that the problem can and will be "solved". In this respect there is no doubt but that any long-term solution to the traffic problems of towns will be associated with significant developments in land-use planning and in transport technology.

Land-use

Transport planning just cannot be dealt with in isolation from land-use planning. While this, of course, is most relevant to planning in large urban/conurbation areas, it is also true with regard to the smaller towns and villages.

The inter-dependence of land-use and transport, is very simply illustrated in Figure 9. This figure is intended to illustrate that whatever the land is used for, its activities generate trips—and some activities obviously generate more trips than others (1–2); these trips, in turn, point up the need for particular types of transport facilities (2–3–4); the extent to which the transport facilities are able to cope with the trip demands determines the quality or degree of accessibility associated with the land in question (4–5); of course, the accessibility associated with the land influences its value since, logically, the land has no value if people cannot get to it (5–6); and finally, it is the land value which helps to determine the use to which the land is put (6–1)—which puts us right back where we started!

Thus, it can be seen that the control of land-use is to a large extent the key to the control of movement, since all trips are generated by land-use. However, initiating *change* in land-use is not something that is easily done under the democratic governmental system, and hence the impact of this solution-approach may not be felt for many years to come; this is particularly so with respect to the planning of the large urban and conurbation areas where, in addition, the pressures are so great that the land-use strategies tend to get overwhelmed by the sheer needs of more immediate "current" problems.

With respect to these great urban areas, it behoves the transportation engineer and planner to keep constantly in mind that while transport planning may be perhaps the single greatest tool for change in the hands of the urban planner, yet the transport demands must be sufficiently flexible that they do not take unwanted precedence over environmental planning—for it is with the environmental aspects

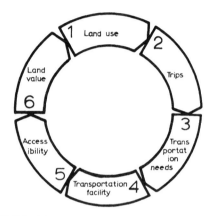

FIGURE 9   The land use–transport cycle.

that most people are mostly concerned. Furthermore, good planning, whatever type it may be, is dynamic planning; it is a continuous balancing process where demands of one nature or another are continually being balanced off against one another. This means that the planning for transport cannot be isolated from planning for houses and urban uses in these great urban areas—indeed, it implies that responsibility for planning as a whole must be a unit responsibility in these areas.

A cursory consideration of the statistics shown in Table VIII is sufficient to suggest that on a national scale there will be at least two generally distinct developments in urban land-use planning. The developments may simply be divided according to whether they will be applied to medium- and small-sized towns or to the great urban areas. From a transport aspect, both approaches will have the common general theme of creating new, and renewing and replacing old, urban areas so that land-uses will be redistributed in time and space; the end aim being that people may more properly enjoy the "quality of life" to which they aspire. The result of this process should be (a) a cutting down on the amounts and lengths of peoples movements, and (b) the creation of definite and improved travelways for particular forms of travel which will not only be separated from each other, but will be designed to service particular designated "environmental" land-uses.

In the smaller- and medium-sized towns, the land-use developments will likely continue along the lines in which they are now generally being guided, i.e. toward the development of the centripetal town of the type idealized in Figure 5. This type has very many advantages from a movement aspect and, practically, it would be undesirable to change to otherwise at this stage. The larger of these towns may require "rapid" public transport facilities which will be separated from the other public and private transport modes.

The most exciting activities with regard to land-use planning will undoubtedly take place with respect to the guiding of development in the great "megalopolis" conurbation-type areas. Current planning thought is turning away from the centripetal concept in relation to these areas, not on the grounds of failure in ideal, but rather on account of the relative inflexibility of their route systems in relation to dealing with unanticipated practical developments in the future. The manner in which transport planning in the conurbations will likely develop has been described by Buchanan et al. (in their study[16] of the region linking Southampton and Portsmouth) as being a "directional grid".

In this study, Buchanan proposed a rational hierarchy of travelways which would accommodate different modes of transport having scales of operation which fall naturally into a graded order. Thus, Buchanan postulated six categories of routes from "1" to "6", each successive route having a larger scale of operation in relation to distance (see Figure 10). The categories proposed would be such that the "1" and "2" routes would correspond to paths and roads in housing areas, while "6" routes would correspond to regional or national communication lines. "1" routes would generally only connect with "2" routes, "2" with "3", and so on; thus at intersections where routes cross would be concentrated the facilities which require accessibility from the two consecutive categories of urban sub-systems served by these routes, e.g. shops would need to be accessible to a residential area at one level and to wholesale distributors at another, or industry would need to be accessible not only to its employees but also to a regional freight route. In Figure 10, the main urban facilities are shown grouped on alternative "red" and "green" routes which Buchanan calls "spines of activity". Thus,

the "red" routes would accommodate public as well as private transport (the two systems running parallel): the "green" routes would be through-routes (possibly some used also by express transport systems) through landscaped areas—they would

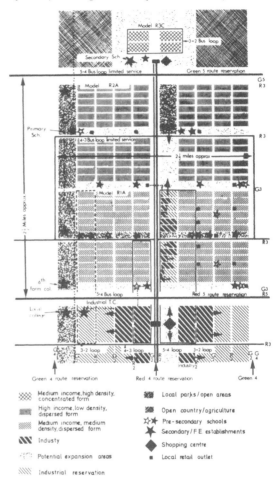

FIGURE 10 Six Routways: a composite diagram showing Buchanan's concept of route structure in a conurbation-type area.[14]

also serve for the random movements which are more likely to be carried out in private transport vehicles.

Transport technology

The development of new transport technology, of new and superior forms of transport to those which now exist, is a necessary part of any "ultimate" solution to the transport/environmental clash in

TABLE X
Effect of transport mode and numbers of workers on the radius of the central area, and on the percentages of the C.A.
devoted to roads and parking

Mode of travel to C.A.	Radius of C.A., mile for a working population of			% of C.A. for carriageways for a working population of			% of C.A. for parking for a working population of		
	10,000	100,000	1,000,000	10,000	100,000	1,000,000	10,000	100,000	1,000,000
Railway	0·11	0·34	0·08	0·2	0·6	2	—	—	—
Bus, wide streets	0·11	0·34	0·11	1	2	7	—	—	—
Bus, narrow streets	0·11	0·35	1·15	1	4	13	—	—	—
Car, wide streets (multi-level parking)	0·12	0·37	1·48	5	16	16	11	11	7
Car, narrow streets (multi-level parking)	0·12	0·42	1·79	9	26	60	11	8	5
Car, wide streets (ground-level parking)	0·17	0·55	1·96	4	11	31	54	51	39
Car, narrow streets (ground-level parking)	0·17	0·57	2·24	6	19	47	54	47	30

urban areas. Certainly any solution which is based on the complete use of the motor car in the larger cities is just not practicable. A measure of the road and parking space requirements which would have to be provided in Central Areas, if this operation was to be attempted, in towns with city centre *working* populations of 1 million (e.g. London, 0·1 million (e.g. Leeds), and 0·001 million (e.g. Walsall)) is given in Table X. (This is the outcome of some theoretical calculations by Smeed[17] as to the consequences for land use if all the commuters to the Central Area used the same modes of transport; the results are based on assumptions which are most applicable to London.) These data show quite clearly (a) that the land requirements for carriageways could rise to prohibitively high levels in the larger towns in particular if all the work trips are to be made by private car, and (b) that parking in the Central Areas in such a situation could only possibly be attempted with the aid of great numbers of multi-storey car parks. Apart from the tremendous financial considerations which would be involved in such an attempt, the magnitude of the constructions required by the all roadway/parking facility approach would be such as to change completely the characters and geographies of the larger urban areas—and this I cannot see being accepted by either the government or the governed of this country.

This, of course, implies that if future workers will not be able to use their private cars to travel to the Central Area then they will either not travel there at all or they will do so primarily via the various public transport facilities. This in turn implies that if the public transport facilities are to be used (since the other alternative is unacceptable) then a "solution"-approach demands that these facilities give a quality of service which will be as nearly compatible as is possible with that provided by the private vehicles which they are intended to replace. In this respect, it certainly cannot be said that the existing public transport facilities give a quality of service which is provided by the motor car—hence the need for new and improved transport techniques.

There are two other points which should be made here with respect to any long-term "solution". Firstly, the transport solution in, for example, a large town will not necessarily be the same as that in a smaller town; in other words, the bigger urban areas will more likely have the large dense concentrations of activities which will lend themselves to the use of particular forms of public transport facilities, whereas the smaller urban areas will not. Secondly, trip patterns in the future will undoubtedly be significantly different from what they are today, e.g. as a result of greater leisure opportunities, so that there will be greater emphasis, for example, on catering for the suburban-to-suburban social and business trip demands as compared with current practice; this also means, of course, that the transport technology which is related to, for example, the future radial public transport trips may not necessarily be the same as that which will cater for the more random cross-town and suburban trips.

"New" transport systems One way by which transport systems may be categorized is according to the degrees of freedom available to them for movement. Thus, the vehicles associated with these systems can be divided into three main groups:

(1) Vehicles that are laterally and vertically restrained
(2) Vehicles that are restrained only vertically
(3) Vehicles with neither vertical nor lateral constraints.

1-Degree of freedom systems Vehicles associated with these systems are free to move only along a fixed line. They, as such, include all tracked and tube vehicles. Typical vehicles are rail and tube vehicles, conveyor belt carriers, and tracked air-cushion vehicles. All such tracked and tube systems, whatever their "newness", are at their most effective when servicing large concentrations of urban activities—which is why so many of them are associated with travel to and in Central Areas. Furthermore, the larger the urban/conurbation area, the more generally suited it is to the use of a tracked system.

It is not possible to describe here all of the systems which have been proposed as ultimate "solutions"; instead brief mention will only be made here of some of the most frequently discussed and sensible ones (for more detail see also, for example, Refs. 18 and 19).

TABLE XI
Urban areas currently considering rapid transport proposals

City	Population (millions) (from standard gazetteers)	City	Population (millions) (from standard gazetteers)
Sao Paulo	5·890	Baghdad	0·913
Calcutta	5·500	Ankara	0·902
Bombay	4·153	Vancouver	0·790
Hong Kong	3·700	Rosario	0·760
Cairo	3·350	Adelaide	0·727
Rio de Janeiro	3·310	Brisbane	0·720
Puerto Rico	2·650	Minsk	0·717
Manchester†	2·500	Bremen	0·706
Sidney‡	2·450	Liverpool	0·705
Pittsburgh	2·376	Dusseldorf	0·704
Delhi	2·360	Antwerp§	0·665
Tehran	2·317	Dortmund	0·651
St. Louis	2·203	New Orleans	0·628
Santiago	2·114	Hanover	0·571
Melbourne	2·110	Auckland	0·515
Singapore	2·000	Helsinki	0·510
Baltimore	1·980	Beirut	0·500
Lima	1·960	Minneapolis	0·483
Madras	1·730	Winnipeg	0·476
Detroit	1·670	Nuremberg	0·466
Bogota	1·490	Perth	0·450
Cincinnati	1·353	Zurich	0·440
Guadalajara	1·300	Hamilton	0·395
Caracas	1·250	Louisville	0·391
Kobe	1·217	Tel Aviv	0·387
Seattle	1·214	Edmonton	0·340
Istanbul	1·152	Charleroi§	0·283
Johannesburg	1·111	Calgary	0·279
Liege§	1·008	Ghent§	0·229
Casablanca	0·965	Newcastle upon Tyne	0·215
Montevideo	0·923	Wellington	0·159

† Traffic area.
‡ Prob. surface line
§ Initially, to be tramway systems.

The most well-known of the tracked systems is, of course, the conventional *Duorail Train*. These vehicles are now in the process of undergoing revolutionary developments with regard to comfort, noise, speed, safety, economy of operation, etc., so that it is not at all wrong to consider future duorails as being part of any future solution-type system. Certainly many cities seem to think so—at this time there are some 80 duorail systems in operation or under consideration in urban areas throughout the world (see Table XI for a list of those proposed).

An indication of the trend with regard to duorails is reflected in the electrified system now near completion in the San Francisco's Bay Area Rapid Transit District (B.A.R.T.D.). This (originally) $792 million, 75 mile long automated rail system is notable for a number of reasons, not the least of which is the fact that it is designed to be in direct competition with the motor car, and hence the system's aim has been to provide such technically advanced high-speed commuting that passengers will be *voluntarily* diverted to it from the motor car. As passengers, on the whole, will not be diverted from car to train unless travel time is saved, the system is designed so that its trains will maintain average schedule speeds of 45–50 m.p.h., the average station stop-time will be 20 sec, and the trains will be capable of achieving maximum speeds of 80 m.p.h. As frequency of service is essential in meeting the convenience requirements of passengers, the interval between trains during the peak periods will be 90 sec, and it will never be greater than 15–20 min during the rest of the 24-h day.

A major difficulty associated with any fixed rail system is that it cannot collect/deposit people at their own doorsteps. To compensate for this the rapid transport operator likes to run his trains at high speeds; this, of course, not only decreases station-to-station time for the passenger, but it also ensures that the transport equipment can make more trips in a given time period. However, the passenger is primarily interested in door-to-door travel time; hence, he is concerned with access time with respect to both the initial and terminal stations as well as the station-to-station time—or, in other words, the average traveller wants the stations to be as frequent as possible so as to minimize his access time. Thus, the duorail system designer seeks an optimum station spacing (see Figure 11) which minimizes total passenger journey time, while at the same time ensuring a high station-to-station train time. (In London, for example, the average station spacing is about 0·125 mile, and the average

train speed is about 24 m.p.h.; with the BARTD system the average spacing will be 2·3 miles.) To overcome this problem with respect to, for example, travel within large Central Areas, a *SPEEDAWAY High Speed Passenger Transport* system has been developed by the Institut Battelle in association

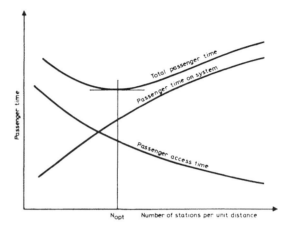

FIGURE 11   The search for the optimum station spacing.

with the Dunlop Company. This system (see Figure 12) uses a series of accelerating treadway sections to enable passengers to load a fast constant-speed continuously moving belt. The key to this system is the boarding treadway which accelerates the passenger at a constant rate along a parobolic path so that the passenger eventually arrives parallel to the main belt, travelling at the same speed, so that all he or she has to do is to step across on to the continuous belt. Disembarkation is the same in reverse and the passenger eventually leaves the system at the same low speed as he entered (2 m.p.h.). This system could well have a high potential—at any rate the National Research and Development Corporation seem to think so, for it is supporting further investigative work on the system. A full scale prototype of this system is now in operation in the Batelle Institute in Geneva. (Details of the SPEEDAWAY and essentially all other conveyor systems that have been proposed for urban areas are available in the literature.[20])

Another form of rapid transport tracked system about which much has been said at one time or another is the *Monorail*. To the layman, the term monorail often conjures up visions of sleek, clean, light, speedy, modern vehicles streaking over the ground on track supported by ethereal stilts. In

FIGURE 12   Passengers decelerating from the SPEEDAWAY passenger conveyor system.

point of fact, this is not the case. First of all, the monorail is not a new vehicle—the first patent was taken out in 1821. Furthermore, one of the only two financially successful commercial monorails was in operation in Co. Kerry, Ireland between 1888 and 1924; it then closed down as a result of road competition. The most financially successful monorail—that built at Wuppertal, Germany in 1898–1903—is still in operation; this is a 9·3-mile long line which operates over the Wupper River for a distance of 6·2 miles, and is supported by A-frames whose foundations are on opposite sides of the river. The newest monorail is the 8·1-mile long line connecting Hamamatsu-cho Station in central Tokyo with Tokyo International Airport. A second point which might be said here is that the newest duorails require no more massive construction supports than the monorails, and hence are generally equivalent with respect to their visual intrusion upon the environment.

Without going into great detail it can be said that, in general, the monorail has little to offer that cannot be matched by new lightweight duorails; indeed it can be said that the reverse is true—particularly with respect to switching and safety arrangements (see, for example, Ref. 21).

The newest and perhaps the bravest of the future urban rapid transport systems is the guided air cushion vehicle or "hovertrain". Although generally thought of as an inter-urban transport system, recent thought and ingenuity has turned attention to its use within urban/conurbation areas. A prototype of such a system—known as the *URBA* system[22,23]—has already been built and demonstrated at Lyon, France. This has a propulsion system that is extremely quiet (68 dB), emits no fumes, and has no moving parts. No vibration is transmitted either to passengers or to the ground, as there is no mechanical contact between the vehicle and the track. URBA is claimed to be capable of carrying the same volume of people as a conventional duorail at similar speeds for lower capital and operating costs. If all this be indeed so, then this system could well make a significant contribution to future rapid transport in large cities, particularly conurbations—particularly from an environmental aspect.

All of the systems mentioned above are essentially mass transport systems capable of moving perhaps 20,000–60,000 people per hour per track at average speeds of between 20 and 45 m.p.h. In recent years, interest has also developed with respect to the use

FIGURE 13   Photo-montage of a 30-seater URBA 30 train.[23]

of semi-personal systems in, particularly, the central areas of very large towns. Typical of these is *Cabtrack*,[24] which is being currently studied and evaluated at the Royal Aircraft Establishment in Britain. As perhaps its name implies, this is a 4-seater tracked taxi capable of moving, under automatic control, about 6000 people per hour (in 4000 cabs per hour) per track in individual self-routing cars moving at 22 m.p.h. While this number is low in comparison with the high density systems noted above, Cabtrack has the advantage that the track would be cheap to construct due to the small size and weight of each cab, and thus it would be possible to build a network providing a much more extensive coverage.

In large Central Areas, there is a need for pleasant and efficient semi-rapid public transport vehicles which can distribute passengers who travel to terminal stations via the conventional rapid transport systems, or which can act as links between shops, offices and the like. Systems such as Cabtrack, which utilize small light vehicles (see Figure 14), have the ability to traverse small-radius curves, and employ cheap overhead structures, and would appear to have qualities which might minimize the intrusive effect on the architectural environment of existing urban areas. From a long-term technical aspect, there would appear to be little preventing the development and, ultimately, the usage of these systems. All that are needed, unfortunately, are large amounts of money, the

willingness to try innovation—and the courage to risk failure.

2-Degree-of-freedom systems   Vehicles associated with 2-degree-of-freedom systems can move laterally as well as along a line. The two vehicles most well-known for this are, of course, the bus and the motor car.

Looking ahead, it is unlikely that the concept of the 2-degree-of-freedom capsule will change—in other words, *some form* of bus and car will likely contribute to any ultimate "solution" in urban areas. These future buses and cars will, however, not only be pollutant and noise free, but they will also be more comfortable, safer and quicker—for they will eventually be capable of being automated at appropriate locations.

What may well be a crude prototype of the future (long-term) private capsule is the *StaRRcar* (see Figure 15). This vehicle would operate in the following manner. In, say, the morning the driver would get into a StaRRcar at his home and drive it (using its own electrically-powered motor) to a major road where it would come immediately under automatic control; the car would then travel into town at about 60 m.p.h. After the vehicle has arrived close to its destination, it would be released from the automatic control and be (manually) driven to a parking station and left there until required later.

The proposed *Automated Bus* is simply a variation of the automated car in that it would be driven

FIGURE 14   A model of a single tract Cabtrack running down Victoria Street, London. Note the cabstop with deceleration and acceleration lanes.

under manual control on minor roads, after which it would come under automated control when it came on the main roads or on reserved guideways.

It might be mentioned here that numerous versions of the automatic car and the automatic bus have been proposed. Generally, however, they are all similar in principal of operation to the systems described above.

One other example of a 2-degree-of-freedom system which should be described here, because of its immense possibilities with regard to the not-too-far-distant future, is the *Dial-A-Bus* system.[25] This system which, as proposed, is essentially a hybrid between a conventional bus and a conventional taxi, would operate in the following way. A customer desiring service would telephone a central control computer from either his home or a bus stop, and would indicate his desired trip starting time, origin and destination. The computer would then decide which bus already in motion could best suit this particular demand; the driver of that bus would then be directly contacted by means of a digital printer communicatons device or by radio

phone, and he would be instructed where and when to pick up the passenger; the passenger, meanwhile would be told at what time the bus is scheduled to arrive at his home or the nearest bus stop, as the case may be. The buses used for this purpose would be "mini-type" buses, i.e. capable of carrying between 10 and 20 people. It is estimated that the cost of a typical trip via Dial-A-Bus would probably be about 50% more than that of a normal suburban bus trip of the same length at the present time.

The proposed Dial-A-Bus system is attractive for a number of reasons. Firstly, of course, it is a demand-responsive public transport system which would give near private transport service, and hence represents an attractive alternative—as well as being a very genuine competitor—to the private vehicle in many circumstances. Secondly, it is a system which is aimed specifically at serving suburbia and small towns, where conventional bus systems, which must follow predetermined routes according to predetermined schedules, tend to come to grief through lack of patronage. It can also, of course, be used to provide "owl" service

FIGURE 15   The "automated" StaRRcar.

at night, when conventional public transport tends to be inefficient. Lastly, but far from being least, it is a public transport system which would bring near-private transport mobility to the oft-forgotten section of the population—the old, the poor, the young, the handicapped, and the non-drivers. For this last reason, if no other, it is a system which is deserving of considerable research and governmental support. It is a system which, in my opinion, should become accepted, even if it loses money—for it would perform a much needed social service.

There is one further system which should be noted here because of its possible influence on ownership—and, hence, the total number of cars on the road. This is the drive-yourself taxi experiment now being promoted in Montpellier, southern France. The scheme operates on a co-operative basis whereby each participant owns shares in a company which owns some 150 cars. Each "owner" is issued with a car key, which fits all vehicles, as well as a set of plastic tokens, each costing 5 francs apiece; when a plastic token is inserted in a meter, the "rented" car can be driven for a distance of 5 miles, after which a new token must be used. The system is intended to alleviate the problem of driver-owned vehicles remaining outside homes and offices when not being used, for the rented car must be left at one of a number of special parks at the end

of a trip so that it can be utilized by somebody else.

3-Degree-of-freedom systems   The most obvious vehicle to be considered here in relation to any eventual "solution" is the airplane. Nevertheless, no matter how fanciful one wishes to get with regard to the future of air travel (e.g. see Ref. 26), it is doubtful whether any solution-type approach to tackling the transport problems of towns will see reliance being placed on the airplane for this purpose. In fact, the mind rather blanches at the thought of having the current transport problems of congestion, safety, noise, etc., being transferred from the ground to the air!

What should perhaps be mentioned here, however, is the possibility that large urban areas could see *VTOL* (*Vertical Take-Off and Landing*) and *STOL* (*Short Take-Off and Landing*) *aircraft* within their boundaries in the late 1980's/early 1990's. This likelihood arises as a result of the "transport gap" which exists with respect to distances in the range 250–750 miles. Unless ground access times to airports can be reduced significantly, the airlines' wish to invade this short-haul high density travel market will not be fulfilled—and many people feel that the simplest way to reduce access time is to negate the need for people to spend long travelling times to get from, say, their homes to the

TABLE XII
Comparison of hypothetical trips by conventional and vertical take-off and
landing aircraft[25]

	London–Paris via		London–Leeds via	
	CTOL	VTOL	CTOL	VTOL
Total door-to-door time (min)	219	150	175	112
Air trip share of total (%)	25	37	34	54
Ave. journey speed (m.p.h.)	55	80	69	107

airport and, instead, to bring the airports into suburbia and use VTOL/STOL aircraft to service them. What this might mean in terms of shortening the door-to-door air trip between, for example, London–Paris and London–Leeds is shown in Table XII.

Before governmental go-ahead is given for passenger-carrying commercial VTOL/STOL aircraft to be developed, it is my belief that there is one very pertinent question which should be asked— "Do we wish to have them in our towns?". The answer may well be yes, particularly if our design engineers and metroport planners are able to guarantee towns that there will be no detriment to the environment associated with their introduction. However, as of yet no such inquiry has been initiated—and no such guarantee is proven—and it may well be that the fact of the situation will be upon us unless some such inquiry is made soon. In this respect we have much to learn from the past— which is that only too often have important events with regard to transport got out of control, simply because critical decisions were taken without sufficient thought being given to their possible consequences.

It may perhaps be stretching the definition a little but another form of a 3-degree-of-freedom "transport" system which the long-term future will undoubtedly see is the "No-trip" transport system. In other words, the rate at which communication technology is developing, and will continue to develop, is such that many trips which currently have to be "made" in person will instead be made in the future by electronic means, e.g. shopping trips will be made by the videophone, social visits will be made via the personalized telephone, certain business trips will be cut down because of the instantaneous transmission of documents. In theory, these developments should greatly reduce the *need* for many trips in urban areas; however, whether or not they will cause an actual reduction in the total number of trips cannot be said at this time —it may well be, indeed, that (as with the telephone) the end result of the introduction of these electronic aids is an increase in the number of trips, simply because of the ease with which these aids enable preliminary contacts to be established.

PEOPLE, TRANSPORT SYSTEMS AND THE URBAN SCENE: AN OVERVIEW—II

C. A. O'FLAHERTY

Institute of Transport Studies, University of Leeds,
Leeds LS2 QJ7, U.K.

This paper, which is divided into two parts, examines the problem of transport in towns, with particular relevance to the relationship between the motor vehicle and urban environment.

This, the second part, places particular stress on transport and land use developments likely in the near future, and the manner in which they can best be encouraged so as to improve the urban environment. The key to the reconciliation between transport needs and environmental ones is considered to be "separation". Suggestions are made with regard to times and locations at which "separation" procedures can be utilized so that movement in the urban area can be more desirably related to environmental considerations.

THE IMMEDIATE FUTURE

From an environmental aspect, there is no doubt but that the ultimate "solution" to the transport problem should also result in the elimination of the detrimental features associated with it, e.g. if transport goes automated then, of course, congestion should be capable of being minimized, while noise, accidents, air pollution, etc. should be essentially eliminated. Unfortunately, however, "solution-day" is still very very far away, so that the questions which must now be considered are (a) what is likely to happen in the near future, within, say, the next 25–30 years, and (b) what basic principles relating transport to the environment should be kept in mind and, where possible, followed during this time.

Perhaps the best way to attempt to answer these questions is to consider again the detrimental features associated with the motor vehicle in the urban environment—only this time the emphasis will be placed on constructive suggestions regarding the means by which these detriments can be minimized or eliminated entirely.

Accidents

In 1968, 254,478 people were injured on roads in built-up areas in Great Britain; 61,713 of these were killed or seriously injured. Of these the following were the pedestrian casualties:

	Killed and seriously injured	All casualties
Children	10,775	37,894
Adults	13,987	40,950
Total	24,762	78,844

There is one simple way in which these pedestrian accidents can be not just reduced, but eliminated—pedestrians and road traffic should be *separated*. For whenever a pedestrian and a vehicle collide, the pedestrian comes off the worst.

Wherever there are large congregations of pedestrians, and particularly at locations which have high accident rates, considerable thought should be given to the possibility of banning traffic entirely. One of the most pleasant features which one now sees being more and more accepted in towns is that of turning large sections of central areas into pedestrian precincts. The primary factor influencing this development is that local authorities and civic leaders are now appreciating that it is necessary for the central area to be made more attractive to shoppers and people on business if it is to survive in the face of competition from rapidly growing suburban and out-of-town shopping centres. This should, of course, have a most desirable by-product with respect to central area pedestrian–vehicle conflicts.

If it is not possible to close-off a particular road to traffic, then it may be possible to divert a major part of the traffic from that road on to one which can more suitably carry it. This can best be illustrated by giving an example. In Leeds at this time, one of the most heavily trafficed radial roads is Woodhouse Lane; every morning and afternoon traffic on this road slows to a crawl because of the numbers of vehicles which it has to carry. Two miles of this road are flanked by an educational complex of schools, the Polytechnic and the University. Recently part of this section (past the University) was made one-way, which immediately reduced the volume of vehicles on the road adjacent to that part of the University very significantly every morning. Now the City of Leeds proposes to construct a new relief road which will have the practical effect of not only improving traffic movement but it will also turn a large part of this existing radial road into a local service road which can be more safely used by school children, students and other pedestrians.

While on the subject of schools, why is it that children must go to school at the same time as the peak traffic goes to work? I realise that many reasons regarding convenience, custom, etc. can be given—but when one weighs these excuses against the opportunities for children walking to school to be involved in accidents, one wonders, as the saying goes, "whether we have our priorities right". Surely children going to school should be separated in time from vehicles going to work.

Nowadays very many of the main roads are so crowded during particular times of the day that motorists leave them and find their way to their destinations via complicated networks of residential streets. Unfortunately, many of these tortuous routes now tend to go through housing areas/estates whose road systems are not at all designed to carry high speed heavy volumes of traffic. Obviously this is undesirable and so every effort should be made to prevent this—if necessary by blocking-off particular streets, so as to keep these unwanted vehicles out of these pedestrian areas.

While on the subject of housing estates, I am quite sure that something needs to be done to reduce vehicle speeds in housing estates. I do not have any figures to substantiate this but it seems to me from visual observation that very many of the cars travelling in what are obviously housing areas do so at speeds which are in excess of 30 m.p.h. when traffic is light. Surely, a more desirable speed should be of the order of 10–15 m.p.h. I do

not have any magic answer as to how to get drivers to reduce their speeds in housing estates—unless it is by imposing severe speed limits; however, I see little value in imposing strict speed limits unless they are going to be enforced in law.

Wherever and whenever it is possible, pedestrians crossing the road should be separated from traffic either in time or in space. *Time separation* implies that people should be encouraged and helped to cross at traffic lights. *Space separation* implies that people should be able to cross the road via a pedestrian bridge or subway wherever possible. In either case, guard rails, which can prevent people from stepping on the carriageway, should be used to channel people to these separated crossings.

Even though there are no pedestrians on particular roadways, accidents will still, of course, occur. Only this time there will be crashes between vehicles and obstructions at the edge of the carriageway or between vehicles themselves on the carriageway. If these accidents cannot be eliminated, much can nevertheless be done to minimize their severity.

For example, many accidents occur between vehicles and roadside poles, posts and suchlike rigid objects. Why must so many of these fixed obstructions be placed so close to the edges of the carriageway—particularly those of high speed urban/suburban roads? I am quite sure that many of these "pillars" could be either eliminated entirely or else placed outside the likely trajectories of crashing vehicles. (This would also, of course, improve the visual amenity of the area in question.) If this is not possible, then why cannot poles be used which yield to the vehicle when impact takes place? There is no reason why these posts should be made of concrete or otherwise rigidly fixed at the side of the carriageway. "Giveway" poles have been designed, and are now in use in the United States which are practical, effective and competitive in cost with currently used typical rigid poles—and these can ensure that a death from a pole–vehicle collision cannot occur.

Another way of reducing the number of street accidents is to eliminate kerb parking. Kerb parking is dangerous because it restricts visibility at street corners (causing junction accidents), and creates pedestrian accidents by forcing people to enter and leave their parked cars on the carriageway. (In the State of Connecticut, 17% of the urban area accidents were recently estimated to involve vehicles which were parked, manoeuvring into or out of a parked position, or stopped in traffic.[28]) In addition, of course, kerb parking reduces street capacity, hampers road maintenance, and restricts the movements of emergency vehicles. Accidents are more likely to occur with angled kerb parking, than with parallel parking for two reasons: (a) drivers tend to steer clear of the rear of angle-parked vehicles and so may intrude into the opposing traffic lane; (b) drivers sometimes will make U-turns to enter spaces on the opposite side. Certainly kerb parking, particularly angled-kerb parking, should not be permitted on heavily travelled urban roads, the footways of which are extensively used by pedestrians.

Why cannot vehicle-to-vehicle collisions be made "safer" by improving the safety qualities of the vehicles themselves? For example, one very easy way of reducing the severity of very many accidents would be to legislate that vehicles in Great Britain should be equipped with collapsible steering wheels, instead of the currently-used solid spear-like steering shafts which point directly toward the driver's chest. Another very simple way of reducing the severity of vehicle-to-vehicle accidents would be to ensure that every car is fitted with effective energy-absorbing front and rear bumpers instead of the pretty ornaments which are currently used on the great majority of British and, indeed, European cars.

There are of course, many other suggestions which may be made to improve car safety so that when accidents do occur (see for example, Ref. 29) the damage to vehicles and their passengers will be minimized. All of these will add extra costs on to the prices of cars—of this there is no doubt—and this could possibly mean that the growth in the numbers of cars sold every year may be slowed down. However, this would seem to be a small price to pay for the many lives which will be saved as a result, and the much unnecessary hardship which will be prevented.

Air Pollution

The answer to the air pollution problem associated with the car is, of course, to control the source, i.e. to use an engine in the vehicle which is other than an internal combustion engine, e.g. a vehicle which is electrically powered or powered by external combustion engines such as the steam engine or the hot air Stirling engine. Without going into detail, however, it would appear at this time that this development is not likely to make any significant dent in the air pollution problem within at least the next 15 years. Instead, therefore, the emphasis will have to be placed on minimizing the undesirable effluents from the existing conventional internal combustion engined vehicle (e.g. by replacing carburettors with fuel injection systems).

The greatest strides are taking place in this respect in the United States. (British cars sold in the U.S. are fitted with special equipment costing about £40 to enable them to meet the current required standards.) In the U.S., new cars are only permitted (by law) to give off less than one-fifth of the exhaust emissions of carbon monoxide, hydrocarbons, and oxides of nitrogen, which were allowed in 1969. Furthermore, it is likely that still higher standards will be set in the near future in America as a result of the research now being conducted there on engines which will run on lead-free petrol.

In Great Britain, the air pollution problem is not taken as seriously as it is in the United States. The main concession being made in this country is contained in the announcement[30] of the government's intention to lay statutory regulations which will require all future new vehicle engines to be fitted with a new "breathing" device which will feed the crank case emissions back into the air intake to the cylinders; this will reduce hydrocarbon emissions by about 25–30%. A new British Standard is also being prepared on the rating of diesel engines which will specify maximum levels of smoke emission from heavy vehicles. Other than these, no further official criteria are being issued at the governmental level regarding emissions from vehicles.

If the traffic planner/engineer can do nothing about controlling the source of pollution, he can do something about ensuring that its effect upon pedestrians in urban areas is reduced at locations where new or reconstructed roads are being built. Again, as with accidents, the planner's key is *separation*, e.g. he can create traffic-free pedestrian precincts in towns, make sure that footways are separated from flanking traffic lanes by grass or adorned concrete or bituminous strips, employ subways or pedestrian bridges at locations where heavy volumes of people wish to cross the road, and use wide traffic lanes wherever possible. Furthermore, when selecting sites for new routes and for large parking facilities, consideration should be given by the planner to the existing wind patterns so as to ensure that a build-up of polluted air is not likely to take place at any given footway or carriageway location.

Lastly, but far from being least, the transportation planner/engineer can reduce the level of atmospheric pollution from the motor vehicle by employing traffic management measures which ensure that traffic moves smoothly and relatively swiftly in urban areas. Note, in this respect that there is typically a four-fold decrease in pollution when traffic speed is doubled from 7·5 to 15 m.p.h., and a six-fold decrease when it is raised from near zero, as in a traffic jam, to 15 m.p.h.[7]

Noise

The best way, of course, to reduce noise is—as with air pollution—to control the source. In this respect, the government in this country is taking very significant steps to reduce the noises emitted from individual motor vehicles. Thus the 1970 maximum permitted noise level for lorries in Great Britain is 89 dB(A), for motor cycles 86 dB(A), and for cars 84 dB(A); the long-term aim[30] is to reduce the levels for lorries to 80 dB(A) and for cars to 75 dB(A). (In judging the effect of this, it should be noted that a reduction of 9 or 10 dB(A) is equivalent to a reduction of one-half in the noise which people subjectively experience.) The government also intends to include a noise check in the present annual check on heavy vehicle maintenance; the aim in this case is to ensure that bad maintenance does not allow a vehicle to become more noisy as it ages.

Despite the above legislation in Great Britain, the fact is that traffic noise is going to continue to be one of the main detriments of the urban environment within the foreseeable future. What can be done to minimize these detrimental effects? If, for example, the daytime noise level generated by an at-grade urban motorway is taken to be about 83 dB(A), and if the acceptable maximum internal noise level in a dwelling is reckoned to be, say, 45 dB(A), then alternative ways[31] of meeting this criterion are (in descending order of *building* cost):

(1) Double glazing and artificial ventilation (at between £1000 and £1500 extra per dwelling).
(2) Specially designed dwellings facing away from the motorway (at £500 extra per dwelling).
(3) Normal dwellings 240 m away with soft landscape in between.

Of course, the most desirable solution is that listed last above—but it is rarely possible in an urban area because of the land cost involved. More practicable is the concept of utilizing as wide as is possible *separation strip* within which would be inserted

trees, bushes, or baffles which would be specially designed to absorb the sound waves (note that densely planted trees and shrubs, on their own, will only decrease noise levels by 5–10 dB(A) per 100 ft of width). Most desirable of all is the concept of separation as outlined above, in conjunction with depressed roadways; not only is this desirable from a noise aspect, but the design can always be made such that it also improves the pleasantness and general amenity of the areas through which the roadways pass.

Most practicable is the reduction of noise at source by introducing improved vehicle design which will reduce aerodynamic noise, door slamming, etc., and improving road design, e.g. by avoiding steep gradients, using road surfacing materials which will reduce tyre noise, etc. Noise levels at particular locations can be also altered by traffic management schemes which will transfer vehicles from one route to another, or give vehicles freer runs and thus minimize the numbers of stops and starts.

Visual Intrusion

There are two aspects of the visual intrusion problem which should be considered. First of all, there is the problem of what to do with respect to the facility on which the vehicles run and, secondly, there is the question of what to do about the vehicles when they are stopped.

With regard to the facility, the most important consideration is that the road and its adjacent land uses must be conceived together as one entity.[31] It is only in this way that the roadway can even attempt to match its environment, and that desired answers can be given to such questions as (a) can noise and air pollution be contained so that standards in adjoining areas are not lowered unacceptably, or (b) are pedestrian accidents likely to be increased, or (c) will the social or pedestrian communications which are essential to the life of the local community be severed, or (d) will the road be visually intrusive, obstruct important views and relate badly with adjacent buildings or (e) does the road form allow the integration in it (i.e. under or over it) of other land uses.

There is one further special point which may be made here with regard to urban motorways. Britain is just now beginning to enter the urban motorway age, and so it has a unique opportunity to construct these facilities so as to enhance, rather than disturb, the urban scene. These motorways represent the most expensive public works programmes that very many cities will see for some time to come, and hence every opportunity should be taken at the planning stage (by taking locational opportunities) and at the design stage (by taking advantage of the cosmetic potentials of landscaping) to ensure that good-looking as well as functional highways are built.

With regard to what to do about the parked vehicle, the ideal answer is again the *separation* of the vehicle from its environment. In certain instances this means the complete prohibition of both moving and standing vehicles from locations where they might clash with the surroundings. If allowed into these areas, the cars should be tucked away in off-street parking facilities (ideally, in multi-storey or underground garages). Where surface parking is or must be permitted, however, as much attention as possible should be paid to the siting and landscaping of the parking facilities so that the standing vehicles fit into the environment and do not stand out drearily against it.

It is also important that properly designed and defined pedestrian walkways should be located so as to give a high quality of service within and into/out of off-street car parking facilities. This is a part of otherwise good car park planning and design which is often forgotten. Furthermore, these facilities should be well lit in hours of darkness, if they are to appear attractive to potential users.

Accessibility and Land Use

Accessibility and land use *can* not only be discussed together but they *should* be discussed together in relation to planning in urban areas. Obvious though this may appear to be, it is nevertheless true to say that it is only in recent years that this has begun to be an accepted concept in relation to towns and their traffic.

The immediate problems associated with accessibility/land use are most obvious in and about town centres. In Great Britain, in particular, the typical centripetal arrangement which is associated with most towns means that it becomes relatively more difficult to obtain access to a Central Area, the closer the motorist gets to it. The result is that, as mentioned previously, town centres now feel that they are fighting for their very survival, as they never had to do so before. In this respect, the following are the transport trends which are/should be developing.

Pedestrian precincts   As mentioned previously, the concept of *separating* pedestrians and vehicles is now becoming more and more accepted. For pedestrianization to be both accepted and successful, however, it must be guaranteed a high quality of access by means of the private car and public transport. Where this has been ensured, the tendency is generally considered to be that pedestrianization causes shopping and commercial activities to be improved within the precinct zone.

The concept of a pedestrian precinct is not, of course, a new one—thus many examples of historical cities which have utilized this form of town centre can be given. The modern development of the pedestrian precinct may be (at least initially) associated with the rebuilding which took place in devastated cities following World War 2. Probably the most well known of these precincts is the "Lijnbaan" shopping centre which was created right in the heart of Rotterdam (see Figure 1a) in 1950. Unlike the shopping street in Leeds (see Figure 1b) where the pedestrian way must still be used by service traffic, the service roads to the shops and businesses in the Lijnbaan are *completely separated* from the pedestrian walkways. Unfortunately, complete separation is generally only possible in specially designed new (e.g. Cumbernauld) or reconstructed (e.g. Coventry) town centres, because of the need for access roads to service buildings within the precinct; thus, pedestrianization developments in this country will in many instances likely take the form of *partial separation* where the pedestrian will have to share the precinct area with delivery/service vehicles and, possibly, public transport vehicles (as is now the case in Leeds).

Road pricing, parking, and the modal split   Whether or not a Central Area has a pedestrian precinct, it must still be well serviced with feeding transport facilities if it is to maintain its economic health. In the extreme, there are three possibilities open to the transportation planner in relation to the provision of transport facilities which will service town centres:

(1) Do nothing and let the congestion problem sort itself out.
(2) Provide for the motor car in its entirety, and let the function of public transport be to carry the remaining travellers to the city centre.
(3) Restrict the movement of the private car in and about the Central Area, and place the legislative, economic and social emphasis upon movement via public transport.

Of the above three approaches the first—the *Do-Nothing* formula—is completely unacceptable. It is entirely detrimental from business, social and environmental aspects. It can only result in the eventual complete destruction of the Central Area as a trading and community centre.

The second approach—the *Provide All for the Motor Car* formula—is that which has generally been adopted in the United States. In that country, there is no doubt but that the motor car decisively dominates economic and social life, and road traffic is catered for to as great an extent as possible, and is only regulated for very compelling reasons. The American approach can, and has been, criticized on many accounts—notably, on the grounds that it has contributed significantly towards destroying the environmental quality of the Central Area, while failing to improve its economic health and well-being (see, for example Table I). The data in Table I (although now somewhat out-of-date) still illustrate clearly that the Central Areas of very many large American cities have generally lost a great deal of their former strength as retail centres, both in absolute terms and in relation to retail sales in the large metropolitan areas of which they are a part. What is perhaps

TABLE I
Per cent change in retail sales of central and metropolitan areas in the U.S. between 1954 and 1963 (reported in Ref. 32)

City	% change in retail sales of	
	Central area	Metropolitan area
Atlanta	+14·7	+63·4
Baltimore	−25·0	+40·0
Chicago	−4·7	+18·2
Dallas	−21·0	+60·1
Detroit	−27·7	+24·8
Houston	−4·0	+58·9
Los Angeles	−17·6	+58·8
Philadelphia	−6·4	+28·3
Pittsburgh	+0·6	+24·2
Seattle	+6·8	+68·4

most impressive about these data is that the trend which they exhibit has occurred in spite of the fact that most U.S. cities have only really grown up in the era of the motor car, and are better adapted by their layout to the requirements of motor traffic.

It is generally considered impossible for both financial and social reasons to adopt the U.S.

approach in the older cities of Europe. Hence the trend has been toward acceptance of the third alternative listed above—that of restricting the numbers of motor cars which may use the town centres.

Before discussing developments which are taking place in this respect, however, mention will be briefly made of one of the more imaginative but nevertheless realistic suggestions which have been made in relation to controlling the usage of motor vehicles in British towns. This is that a pricing mechanism be utilized[33] which would act in the nature of a *Congestion Tax* in that vehicle owners would be required to pay higher charges or taxes when they used congested roads, and lower charges when they used non-congested ones. Thus, only vehicles which must travel on congested roads (in, for example, town centres) would likely be willing to pay for the privilege of doing so, while other vehicles would be influenced by the magnitude of the charges involved from making unnecessary trips on the same facilities.

Road pricing (as the above is termed) is currently being very seriously studied by Britain's Road Research Laboratory because it is, in many senses, the ideal way to deal with the road traffic problem in towns and their centres, i.e. it would ensure that the motorist (and not Society) would pay for the facilities provided, the numbers of vehicles to be handled could always be kept down to the desired level since the charges could be so pitched as to regulate the numbers of vehicles using the roadways at given locations, and lastly, but far from being least, it would likely have the end product that direct car taxation could be significantly reduced so that more people could have their own magic carpets, only they would not be able to use them freely in particular areas of town.

Ideal though the road pricing concept may appear to be from environmental, economic and social reasons, it is this writer's opinion that it is unlikely to be adopted in any city in Great Britain in the near future. Technically, it is quite feasible; politically it is not.

Instead of utilizing road pricing as a means of limiting vehicular movements in towns, parking control is now generally considered to be the key to restricting the movements of private cars in Central Areas. Or perhaps a better way to put it would be to say that parking control is the means by which the *Trip-Purpose Separation* formula—the policy of separating wanted from unwanted traffic in the central area—will be implemented.

The term "parking control" as used here refers to the conscious decision which is now generally being taken by local and national authorities in this and some other European countries to restrict traffic movements into and within the central areas of large and medium-sized towns to artificially low levels by deliberately imposing drastic parking restrictions. The restrictions imposed are intended to result in a limiting of the total number of parking spaces provided (so that the feeding road system is kept within its capacity): a limiting of the length of time during which a vehicle may stay in a controlled parking space (so as to keep out the journey-to-work vehicle, e.g. in London in 1966 it was estimated[34] that the cars belonging to 2% of the Central London working population occupied over 25% of the total available parking space—and about half of this 2% parked free in the streets): and the imposition of a relatively severe parking price for the rent of a parking space (so as to

ensure that only those who need to park in the central areas for short periods of time will do so).

What is probably a typical example of the manner in which central area policies are developing in this direction is reflected in the proposals now being implemented in Leeds.[35]

In Leeds, the decision has been taken by the city council to control the total quantity of long-stay parking in the central area, its situation and the charges to be made, so that the number of commuter car journeys which will be made in 1981 will be limited to a number which (allowing for appropriate occupancy rates) sets a maximum limit to the growth in usage of private car transport at approximately 20% of the 163,000 person-journeys to work in the central (business and industrial) area. The decision to settle on a modal split of 20% private car—80% other mode was taken on the grounds that it ensured (a) that the peak hour demand would be within the highway capacity which could

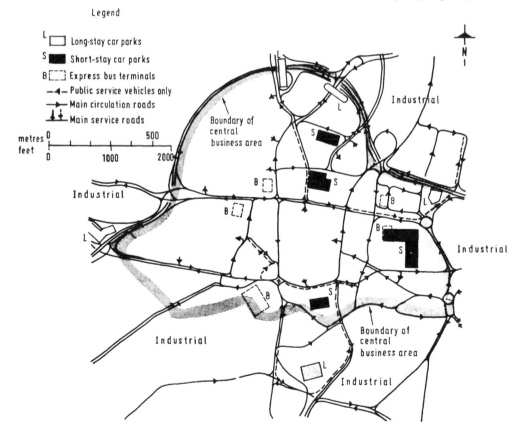

FIGURE 2 The Leeds central area traffic control plan for 1981.

be provided in a realistic plan for the design year, and (b) that the remaining demand—which would be primarily (66%) for bus transport—would be sufficient to support an efficient public transport service at satisfactory frequencies.

In addition to the above long-stay parking spaces, approximately 8000 on- and off-street car parking spaces are to be provided for short-stay shopper and business parking needs, i.e. for up to 2·5 h. within the central area of Leeds. These will be close to offices, shops and warehouses and linked with the special pedestrian ways which at the time of writing are being created within the city centre.

The general outline of much of the Leeds plan is reflected in Figure 2. Note, first of all, that the road system—part of which is an urban motorway—is designed to enable traffic which does not wish to enter the central area to by-pass it. The long-stay multi-storey parking facilities are all located at the fringe of the central area so as to intercept the commuter traffic; the intention is that the commuters will leave their cars at these facilities when they come into town, and then either travel on foot or by public transport to their work destinations. In contrast, the short-stay parking facilities and the public transport terminals are generally located well within the central area, thereby ensuring that the users of these facilities are close to their destinations.

As a complement to the above proposals, Leeds City Council plan to develop the following new forms of passenger road service in addition to the normal inter- and intra-district bus services:

(1) Express bus services which will use the primary road network and provide non-stop fast services between suburban community centres and the central area.

(2) City centre mini-bus services which will provide for short (shopping and business) movements within the central business area.

(3) Park-an'-Ride Services which will link outer suburbia interchange points with the central (business and industrial) area.

One further point may be made here with regard to the Leeds approach. This is that it also to a certain extent represents a brave attempt to reverse the "transport spiral", i.e. whereby the slowness of public transport on congested roads encouraged people to use their own cars, which in turn, made the roads more congested, which in turn encouraged . . . If the Leeds plan is successful, then the radial roads leading to the central area will be within their capacities at all times, which means

that they will not be congested so that people will be encouraged to leave their cars at home and travel via the buses which will be able to move more swiftly, which will mean that people will be encouraged to leave their cars at home and travel via the buses which will . . .

Park-an'-Ride It is appropriate here to mention Park-an'-Ride as a means of promoting ease of access to particular land uses, and especially to central areas. There are two general forms of Park-an'-Ride, both of which are represented in the Leeds proposals.

The first of these forms is represented graphically in Figure 2: note that the commuter motorists are only permitted to drive their vehicles to the long-stay car parks located at the periphery of the central area; from there they are expected to go by public transport (or walk) to their destinations. This form of Park-an'-Ride will likely prove reasonably successful (from the point of view of the transportation engineer) in most towns, irrespective of whether they be large, medium or small, provided that there is firm control of parking within and adjacent to (outside) the periphery.

The other form of Park-an'-Ride is that whereby motorists drive their cars to long-stay car parks which are located well away from the central area, usually in suburbia, and then travel by public transport to their destinations. This is not a new development; in fact, it grew of its own accord for some considerable time before the traffic engineer seized on it as a means of helping him solve his problems, e.g. many people working in city centres have always driven themselves to favourably-located suburban rail stations and then travelled via rapid public transport to their work destinations. Now this practice has become accepted as a desirable transport principle and is applied to bus public transport as well as to rail transport.

It is difficult to say where the latter type of Park-an'-Ride is likely to be successful in Leeds—indeed it is even difficult to define what is meant by a "successful" Park-an'-Ride system anyway. For example, a successful system to a public transport operator is generally one which makes a financial profit; in contrast, the transportation engineer may well consider the Park-an'-Ride system to be successful if it takes significant numbers of vehicles off the roads leading into the central area and, in so doing, helps obviate the parking problem—even though the system may not be self-supporting in the process. Irrespective of which of these definitions is preferred, this form of Park-an'-Ride is much more

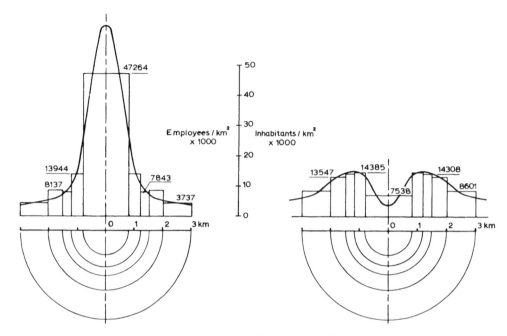

FIGURE 3   Division between the working population (left) and the resident population (right) in Hanover.[36]

likely to approach success in a large town rather than in a smaller town. Why this should be so is suggested in Figure 3 which represents the division between the working and resident populations in Hanover, W. Germany. The diagrams in this figure first of all reflect the fact that while very many workers may have their places of employment in a central area, very few will live there. Note, however, that the concentration of residences builds up as the distance from the central area is increased until a maximum is reached, after which the concentration decreases. Now, in any town in which it is wished to initiate a car–bus Park-an'-Ride system from suburbia, the location of the Park-an'-Ride interchange is going to have to be a significant distance away from the city centre or otherwise it will just not be worth people's while to use the system. On the other hand, if the town is not a large one then an interchange located the same distance away from the central area may just not be able to generate enough traffic to justify itself. For obvious reasons, therefore, it can be seen that the likelihood of there being enough passengers to sustain such a Park-an'-Ride system becomes greater the larger the population of the town.

A basic requirement of any Park-an'-Ride scheme is, of course, a parking facility where vehicles can be left all day while their drivers and passengers are in the central area. These car parks should also contain specifically outlined (adjacent to the public transport stop, if possible) *Kiss-an'-Ride* spaces where members of the commuters' families can briefly park their cars while dropping in the morning or awaiting in the evening, the travellers.

Kiss-an'-Ride is the oft-forgotten part of most Park-an'-Ride programmes. It poses its most severe problems where an express bus (or rail) service is the medium of transport, and buses (or trains) operate with short headways. The dropping of passengers in the morning need not be a serious problem; the crunch comes, however, in the evening when cars parked in the surrounding streets or on the internal roadways of the parking facility itself become a cause of serious traffic congestion. I might say here that it is useless to attempt to stop this practice by legal means— simply because it is essentially unstoppable. What policeman or traffic warden with a family of his or her own is going to risk arresting or giving tickets to bevies of harassed housewives at the wheels of cars full of starry-eyed or rampaging children (as the case may be) while they wait for the bread-winner to return home from work. A much better answer is for the transportation planner to provide short-term parking spaces for these vehicles.

Priority for public transport Most transportation planners/engineers will hold the view that it is just not good enough to require people to leave their cars at home or at, for example, a suburban parking facility, and then expect them to travel into the central area in a bus which is continually held up by other vehicles. Instead, they generally feel that a high quality bus service should be provided which will compensate the motorist to a certain extent for giving up his "magic carpet" temporarily. It is for this reason that consideration is now, more and more, being given to the use of express buses to provide high quality services for public transport users.

Unfortunately, to many local authorities the term "express bus" simply means a bus which makes less stops than before between terminals. While this type of operation can, and obviously will, cut down on bus travelling time, the public transport vehicle is still subject to the congestive influence of the heavy volumes of motor cars present in the traffic stream. What is much more desirable is that the public transport vehicles should be separated from private cars while making their express runs. Ideally, this should mean *Complete Separation*, i.e. on separate routeways such as are indicated in Figure 5b (Part I). Because of the high cost of complete separation, however, it is unlikely that this will be accepted in the near future in any but a very few large towns (and these will then probably opt for rapid transport *train* services). What is much more likely to be accepted and implemented is the concept of *Partial Separation* whereby public and private transport vehicles would share the same routeways but in all or particular instances, priority of movement would be given to the public transport vehicles.

Examples of the manner in which partial separation techniques might be applied are as follows:

(1) The creation of reserved lanes along main roads for buses, taxis, police cars, fire engines and ambulances.

Buses operating unhindered in these reserved lanes could move between 25,000 and 30,000 seated passengers per hour per lane at a speed of 35–40 m.p.h. Note in this respect that the average car occupancy is in the order of 1·5–1·9 persons, which gives a capacity of about 3000 persons per hour per lane which perhaps might be taken to imply that a reserved public transport lane should be considered whenever the buses along the route in question carry more than 3000 people per hour per lane.

(2) The separation of public and private transport at road junctions.

This could be a costly procedure in that ideally it would require public and private transport to cross each other at different levels. As such it would probably only be acceptable (initially at any rate) in the very large urban areas. Much more likely to be accepted is the practice whereby vehicles in the reserved public transport lanes would be given time priority over other vehicles (at signal controlled intersections) by activating contact pads which would be placed in the reserved carriageway surfaces.

(3) Priority of movement given to buses leaving kerbside loading points.

Buses leaving kerbside loading points (particularly during peak periods) are constantly being faced with the problem of finding entry points into the traffic stream. Often, also, they are faced with badly parked cars occupying part of the (usually inadequate) length of kerb or lay-by supposedly reserved for buses—or then again it may be that parked delivery vehicles prevent buses from drawing up at the bus stop. These difficulties could largely be overcome if the buses were to be given real priority over other vehicles when stopping at/leaving bus stops. Thus, for example, where a bus stop is located in a lay-by located reasonably near a major junction, then the lay-by should be extended to form a separate bus lane into that junction.

(4) Authority for public transport vehicles to move along a reserved lane in the "wrong" direction along a one-way-street.

This procedure is practised very effectively at this time along a 300-yd stretch of roadway leading from the Central Area of Leeds. Its advantages are obvious.

Lower bus fares It is high time that city councils faced up to the fact that public transport is generally regarded by the general public as being *public*, i.e. that it is part of the social services provided by the city to enable its people to move about. (This, of course, is admitted indirectly every year in very many towns when the city transport operators receive subsidies to cover their losses for the previous year.)

I am not advocating that public transport should be free, but rather that urban areas should get away from the concept that public transport must be *entirely* self-supporting, and that every increase in operating cost must be matched by a corresponding increase in fares. This can only have detrimental effects in the long run. In this respect it is useful to remember that the people most hurt financially by fare increases are the poorer members of the community i.e. those who *must* use public transport. A second, technical and practical, point to note is that high fares do not lend an aura of encouragement to the process of weaning the motorist from private to public transport. This latter point is particularly applicable to express bus services whose costs—in theory at any rate—are less than those of local buses, i.e. the principal costs involved in operating a bus are time items not distance items and an express bus's time costs per passenger carried are lower than those for a local bus since an express bus can make more trips in a given unit of time.

What is often forgotten, unfortunately, is that the primary goal of a public transport service is to provide mobility and not to make a profit. Nowadays, an efficient public transport system is also regarded as being an integral part of most cities' plans to save their central areas from economic ruin, and to help towns as a whole to preserve their environmental values from the onslaught of the motor car. In this respect it would appear more rational that city taxes i.e. money raised through the medium of the rates, should be used (at least partially) for this purpose rather than that bus fares should be continually increased, so that public transport continues to appear less attractive to both potential and existing users.

Night-time deliveries One of the most frustrating features of central area life is the congestion caused by commercial vehicles standing at kerb-sides. This is particularly aggravating in European cities where it is not the custom for town centre commercial establishments to have special loading/unloading facilities for delivery vehicles. On top of this is the fact that by virtue of their sheer bigness, very many of these commercial vehicles are particularly intrusive in many of the old towns of Europe with their narrower streets, historic buildings and generally pleasant surroundings.

One wonders therefore, why these delivery vehicles cannot be *separated* in time from other vehicles. Or, to put it another way, "Why do not central area delivery vehicles operate at night?"

Of course there are many and good reasons as to why they should not. But then again, so many of these same reasons apply to industrial workers, policemen, taxi drivers, computer operators, etc. Surely if they can work at night, why cannot delivery vehicle drivers and some central area shop workers also do so? The answer, of course, to this query depends on whether or not one considers these latter people to belong to the important minority who *must* work during the night.

There are very significant economic and environmental advantages to be gained from night-time deliveries. In particular, vehicle journey speeds would be higher at night-time so that the operating costs of the commercial operator would be lower. With night deliveries to large central area shops and stores (or to co-operative collection facilities in the case of small businesses), the commercial vehicles would be able to readily park at the kerb and the drivers can efficiently and safely go about their business; with day time deliveries, on the other hand, delivery vehicles will continue to obstruct traffic movement, will continue to be potential causes of accidents, and will generally contribute to an economic loss by both public and private transport and to an environmental loss by the community as a whole.

SUMMARY AND CONCLUSIONS

I personally believe that the motor car is one of the great technological miracles of the present-day. It is the modern magic carpet which has brought together and, indeed, made possible comfort and personal mobility. As such, it can genuinely be considered to be a facility which contributes significantly to very many people's high standard of living and, in general, to their getting "more out of life".

A recognition of the very many and varied advantages of the motor car does not, however, necessitate blindness with respect to the detrimental by-products of its development. These detriments are most obviously emphasized in urban areas. They are reflected in the congestion, noise, accidents, air pollution, loss of amenity, etc. which are experienced every day in towns, particularly in and about central areas. It is with methods to alleviate these detriments and, by so doing, to raise the quality of enjoyment of the environment within urban areas—that this paper has been primarily concerned.

Ultimately the "solutions" to the transport problems of towns and, hence, the improvement of the qualities of the urban environment which are associated with transport, must be associated with developments in land-use planning and in transport technology. Unfortunately, however, it would appear that "solutions" of this nature are many long years away yet. Meanwhile, every effort must be made by means of shorter-term planning and management procedures toward ensuring that towns and their transport facilities are as we wish them to be—compatible and complementary.

For the immediate future—say, the next 30 years or so—the available indications are such that it would appear that one word can be used to sum up the basic principle which will have to be followed if transport and the urban environment are not to be the enemies which they so often appear to be today. The guiding word, or principle, is "*Separation*". Its utilization means that

— the main travelways should be separated from the areas where people live
— main trunk routes should be separated from local distributor routes
— motor vehicles should be separated from pedestrians at locations where there is heavy and vulnerable pedestrian activity
— public transport should be separated from private transport along congested routes and at congested locations
— journey-to-work parkers should be treated separately from other central area parkers
— delivery vehicles should be treated separately from other vehicles in town centres.

In addition every effort should be made to improve public transport so that those who can not or will not use private transport will be provided with as near private car quality of movement as is practicable—for that is what the people wish.

REFERENCES

1. F. E. Bruce, "The evolution of public health engineering," *J. Roy. Soc. Arts* **102** (4925), 475–494 (1954).
2. Ministry of Transport, *Passenger Transport in Great Britain*. H.M.S.O., London (1971).
3. W. S. Smith, "Urban transport co-ordination" *Traffic Engineering and Control* **8** (5), 304–306 (1966).
4. British Road Federation, *Basic Road Statistics*. The Federation, London (1970).
5. Report of the Committee on the Problem of Noise *Cmnd. 2056*. H.M.S.O., London (1963).
6. C. G. Bottom and D. J. Groome, "Road traffic noise—its nuisance value" *Appl. Acoustics* **2** (4), 279–296 (1969).
7. Ministry of Transport, *Cars for Cities*. H.M.S.O., London (1967).
8. C. Buchanan *et al.*, *Traffic in Towns*. H.M.S.O., London (1963).
9. R. J. Smeed, "Traffic studies and urban congestion" *J. Transp. Econ.* **2** (1), 33–70 (1968).
10. C. Sharp, *Problems of Urban Passenger Transport*. Leicester University Press, Leicester (1967).
11. A. H. Tulpule, *Forecasts of Vehicles and Traffic in Great Britain: 1969. RRL Report LR288*. The Road Research Laboratory, Crowthorne, Berks (1969).
12. P. Hall, "Transportation" *Urban Studies* **6** (3), 408–435 (1969).
13. The Government Actuary, *Population Projections 1970–2010*. H.M.S.O., London (1971).
14. Registrar General, *Annual Estimates of the Population of England and Wales and of Local Authority Areas, 1969.* H.M.S.O., London (1970).
15. C. D. Buchanan and Partners, *The Conurbations*. The British Road Federation, London (1971).
16. C. D. Buchanan and Partners, *South Hampshire Study: Report on the Feasibility of Major Urban Growth.* H.M.S.O., London (1967).
17. R. J. Smeed, "The road space required for traffic in towns" *Town Planning Rev.* **33**, 279–292 (1962/63).
18. C. A. O'Flaherty, *Passenger Transport: Present and Future.* University of Leeds Press, Leeds (1969).
19. W. Avery, R. A. Makofski and R. C. Rand, "Advanced Urban Transportation Systems: An overview" *Proceedings of the Carnegie-Mellon Conference on Advanced Urban Transportation Systems held at Carnegie-Mellon University, May 25–27, 1970*. The University Transportation Research Institute, Pittsburgh, Pennsylvania (1970).
20. J. M. Tough and C. A. O'Flaherty, *Passenger Conveyors*. I. Allan, Shepperton, Middlesex (1971).
21. D. O. Eisele, "State of the art—urban transportation systems" *Proceedings of the American Society of Civil Engineers, Transportation Engineering Journal* **95** (TE3), 439–461 (1969).
22. M. Barthalon, "The invention and development of a suspended air cushion passenger transport system in France" *The Inventor* **9** (1), 1–13 (1969).
23. M. E. Barthalon, J. Rechou and P. Watson, "URBA in business" *Hovering Craft and Hydrofoil* **8** (8), 3–11 (1969).
24. M. G. Langdon, "The Cabtrack urban transport system" *Traffic Engineering and Control* **12** (12), 634–638 (Apr 1971).
25. P. R. Oxley, "Dial-A-Ride demand actuated public transport" *Traffic Engineering and Control* **12** (3), 146–148 (July 1970).
26. E. A. Ulsamer, "Planning tomorrow's total air transportation" *Aerospace International* **3** (3), 27–29, 32–36 (1967).
27. C. A. O'Flaherty, "The airport and the town" *Proceedings of the 21st Symposium of the Colston Research Society, held at Bristol University, March 24–28, 1969. The Colston Papers* **21**, 173–213 (1970).

28. J. H. , Heyman"Parking trends and recommendations" *Traffic Quart.* **22** (2), 245–257 (1968).
29. W. Haddon, "A logical framework for categorizing highway safety phenomena and activity" *Proceedings of the 10th International Study Week in Traffic and Safety Engineering held at Rotterdam. Sept. 6–12, 1970.* The World Touring and Automobile Association, London (1971).
30. The Protection of the Environment (The Fight Against Pollution) *Cmnd. 4373.* H.M.S.O., London (1970).
31. O. Cox, "Methods of reducing amenity losses caused by traffic-roads" *Proceedings of the 10th International Study Week in Traffic and Safety Engineering held at Rotterdam, Sept. 6–12, 1970.* The World Touring and Automobile Association, London (1971).
32. P. G. Koltnow, "The road capacity of an urban centre: its effect on environment and activities" *Proceedings of the 8th International Study Week in Traffic and Safety Engineering held at Barcelona, Sept. 5–10, 1966.* The World Touring and Automobile Association, London (1971).
33. R. J. Smeed *et al. Road Pricing: The Economic and Technical Possibilities.* H.M.S.O., London (1964).
34. B. V. Martin, "The development of parking policies in London" *Proceedings of the 10th International Study Week in Traffic and Safety Engineering held at Rotterdam, Sept. 6–12, 1970.* The World Touring and Automobile Association, London (1971).
35. Leeds City Council *et al. Planning and Transport—The Leeds Approach.* H.M.S.O., London (1969).
36. F. Lehner, "Conserving the vitality of town centres" *Revue de L'UITP (International Union of Public Transport)* **14** (3), 335–353 (1965).

PART TWO

Safety

TRAFFIC SAFETY: PROBLEMS AND SOLUTIONS

JOHN E. BAERWALD

Highway Traffic Safety Center, University of Illinois, Urbana, Illinois 61801, U.S.A.

The author discusses the three main elements of the highway safety problem—human, vehicle and roadway—with special attention to recent accident information, subsequent research activities, and remedial programs. Both accident occurrence and preventative countermeasures are considered within the framework of a three-phase sequence; precrash, crash, and postcrash phases. Methods for identifying and controlling problem drivers (especially intoxicated drivers), improved "packaging" systems for vehicle occupants, and increasing the crash-worthy qualities of motor vehicles are summarized. The Experimental Safety Vehicle Program is cited as an example of the type of international cooperation that should be expanded with participation by all nations.

All countries are becoming increasingly concerned about the growing magnitude of crashes, injuries, and deaths which have paralleled the ever increasing numbers of motor vehicles and mileage driven. This has caused public and private agencies of many countries to accelerate and expand research projects and remedial programs involving the three main elements of the highway safety problem:

(1) The human, as embodied in the driver and pedestrian; his knowledge, skill, and attitude.
(2) The vehicle; its design, manufacture, and maintenance.
(3) The roadway and its environment; its engineering, construction, maintenance, and control.

The use and resultant impact of motor vehicles varies greatly from country to country because of such diverse factors as population density, volumes and types of traffic, and the degree and nature of land utilization. In mountainous Nepal there are over 2000 persons for every motor vehicle, while the United States has slightly less than two persons per vehicle. The world-wide average is 15 persons per vehicle. Great Britain's roads are the most crowded in the world with 62·6 vehicles per mile, while the United States has 28·6 and India has only 1·7. Table I lists the population per vehicle and the road accident fatality rate per 100 million vehicle-kilometers for selected countries in 1969.

Preliminary estimates of the National Safety Council indicate that in 1971 the United States had 55,000 deaths and 2,000,000 injuries (disabling beyond day of accident) caused by motor vehicles at a cost of $14·3 billion. As indicated in Table II, motor vehicle crashes were the leading cause of

TABLE I
Population per motor vehicle and highway fatalities per 100 million vehicle-kilometers for selected countries, 1969

Country	Persons per vehicle	Persons killed (per 100 million vehicle-kilometers)†
Zambia	42	71
Ivory Coast	52	53
Syria	117	49
Turkey	149	31·4
Jordan	81	30·5
Morocco	53	28·9
Yugoslavia	29	19
Venezuela	15	14
Hong Kong	37	13
Cyprus	8·3	11
Japan	6·8	10
France	3·7	8·5
Netherlands	5·1	7
Lebanon	18	6·4
Finland	6·2	6
Australia	2·7	5·8
Denmark	3·8	5
Norway	4·5	5
New Zealand	2·5	4·4
Canada	2·7	4·2
Great Britain	4·0‡	3·8
United States	1·9	3·3

†Deaths occurring within 30 days of accident except 28 days in New Zealand and 1 year in the United States.
‡Includes Northern Ireland.
Source: "1971 Automobile Facts and Figures" (Automobile Manufacturers Association, Inc., Detroit, Michigan, U.S.A.), and "World Road Statistics 1966–1970" (International Road Federation, Washington, D.C., 1971).

death for American males between the ages of 5 and 34 and females aged 5–9 and 15–29.

TABLE II

Rank of motor vehicle crashes as cause of death by age and sex in the United States, 1968

Age Group	Male Rank in age group	Male Percentage of age group	Female Rank in age group	Female Percentage of age group
Under 5	6	2·4	6	2·6
5–9	1	31·6	1	28·4
10–14	1	38·5	3	21·2
15–19	1	47·4	1	48·6
20–24	1	48·3	1	34·4
25–29	1	33·7	1	22·6
30–34	1	22·2	2	13·0
35–39	2	13·6	3	8·4
40–44	3	8·9	5	5·5
45–49	5	5·4	5	3·7
50–54	6	3·7	7	2·6
55–59	8	2·4	7	2·0
60–64	10	1·6	8	1·5
65 and over	11	0·9	12	0·5
All age groups	4	4·0	9	2·0

Source: *Vital Statistics for the United States, 1968, Volume II, Mortality, Part A* (National Center for Health Statistics, Public Health Service, U.S. Department of Health, Education, and Welfare, Washington, D.C.).

Although there is some variation from country to country, the conditions which contribute most to motor vehicle crashes, are driver errors (including failure to yield right-of-way, following too closely, improper turning, speed too fast or too slow for conditions, improper use of traffic control devices, and excessive use of alcohol); design, construction, and maintenance inadequacies of vehicles; and outmoded and substandard roadway facilities.

HUMAN SAFETY PROBLEMS AND SOLUTIONS

Both accident occurrence and preventive countermeasures may be considered within the framework of a three-phase sequence.

The first or *precrash phase* involves such factors as initial and continuing driver education, periodic driver examinations, existence of bad driving or walking habits and attitudes (inattention, discourtesy, disrespect for authority and the rights of others, etc.), and physical and mental deficiencies—

including those induced by alcohol and other drugs.

The second or *crash phase* involves the interaction of the vehicle or vehicles, their occupants, and/or pedestrians. This phase begins when mechanical forces in excess of those the vehicle occupants and pedestrians can tolerate start to exert themselves on vehicles and people. During this phase, vehicles strike other vehicles, pedestrians, or fixed objects. Vehicle occupants may collide with objects within the vehicle or be ejected from the vehicle.

The third or *postcrash phase* involves the physical damage to humans, vehicles, and other objects which result from the crash. This damage may be temporary or permanent, superficial or extensive.

Many countermeasures or remedial actions have been and are being proposed for each of these phases.

Countermeasures in the *precrash phase* cover a wide range of activities and interests. They include the development of comprehensive driver preparation and reeducation programs. It is also essential that beginning drivers undergo extensive testing of both their mental and physical ability to successfully operate a motor vehicle under most normal and emergency situations. Experienced drivers must be periodically reexamined in order to evaluate their physical and intellectual capability to continue driving.

Instead of merely testing a driver by means of simple memory tests on basic driving rules and checking his depth perception and static visual acuity, much more comprehensive testing techniques should be developed. These tests, many of which can be administered by means of simulation devices, should also evaluate such complex and varied characteristics as mental attitude and aggressiveness, judgmental ability in emergency situations, neuromuscular coordination, and dynamic visual acuity.

Driving simulators are now being developed which enable a single instructor to work simultaneously with several students and give them a more comprehensive learning experience in a shorter period of time than is now possible. Specialized instruction can now be offered to physically handicapped and mentally retarded individuals.

Fundamental to a program of driver improvement is an effective method for identifying "problem" drivers. These are individuals who have a disproportionate involvement in traffic law violations or crashes. This involvement may be the result of mental or physical disabilities and can only be identified through a complete and functioning traffic record system which includes a central driver register and an efficient traffic accident reporting procedure. The California Department of Motor Vehicles, in a study of 14,000 young motorists, found that poor school adjustment, poor academic achievement, high driving mileage, and the number of cigarettes smoked, were factors in poor driver performance.

The United States is currently actively pursuing a program of identifying one special type of problem driver—the drinking driver. In 1970, over half of the 55,000 highway deaths which occurred in the United States involved alcohol. Two-thirds of these 28,000 fatalities involved people with a serious alcohol problem. One-third involved heavy social drinkers or inexperienced youths who on occasion drove while intoxicated.

This high percentage of fatalities attributable to alcohol in the United States is in stark contrast to that of Sweden, where a strict drinking-driving control program has been in effect for years, and where alcohol-connected deaths are between 10 and 12 % of total traffic fatalities.

European nations, particularly the Scandinavian, have had long experience in combating drunk driving. They have found that strict legal measures have had an undeniable effect in limiting the role of alcohol as the cause of traffic accidents.

In Britain, 0·08% of alcohol in your blood while driving will get you an automatic first offense punishment of a fine up to £100 and a 1-year suspension of your driver's license. In Norway, a driver is deemed legally drunk if his blood contains as little as 0·05% alcohol, and that will land him in jail for 3 weeks with a year's suspension of his license. The Danes also say 0·05% is a violation, and in Finland, offenders are put to work at hard labor for 3 weeks. One to two months' imprisonment is a common consequence for drunk driving in Sweden.[1]

Countermeasures in the *crash phase* involve development of improved vehicle designs which will better "package" the occupants of the vehicle while simultaneously making the vehicle more crashworthy.

In 1962, the Automotive Crash Injury Research project of Cornell University (U.S.A.) released a report on the leading causes of injury in automobile accidents based on data on 26,131 occupants of 1956 and later model cars in rural injury-producing accidents.

Of these, 1956 models accounted for 31% of the occupants; 1957, 34%; 1958, 17%; 1959, 13%; 1960, 5%; and 1961, 0·1%. It was found that Instrument Panel was the cause of the greatest number of injuries (the criterion here is *any* injury without regard to severity), followed closely by

Steering Assembly. At a somewhat lower level were Windshield, Door Structures, and Ejection.[2]

These causes accounted for 83% of the total injuries. When the injuries were ranked in accordance with severity (by applying weighting factors of 1 for minor, 2 for nondangerous, 4 for dangerous, and 8 for fatal) and the data further classified by type of impact (front impact, side impact, rear impact, rollover, and other impact), it was found that, "Front impacts account for the majority of injuries caused by Instrument Panel, Steering Assembly, and Windshield. Rollover contributes most to Ejection, while side impact produces most injuries due to Door Structures."[3] Research findings such as these led to the subsequent development and required installation of improved door locks, energy-absorbing steering control systems, and seat belt installations in American cars.

Lap belts and shoulder belts or harnesses are now required standard equipment on all American vehicles. Unfortunately, the rate of usage is about 30% for seat belts and approximately 5% for upper torso belts. A Netherlands study in 1968/1969 showed that:

... outside built-up areas in the Netherlands 22% of the passenger cars had safety belts (installation is not required). Of these, 39% were really used by drivers. Hence, 8·5% of all passenger car drivers interviewed (outside built-up areas) used safety belts.[4]

Occupant restraint systems are classified as either active or passive. An active system such as a lap belt or a shoulder harness, requires some action on the part of the occupant, while the passive systems function automatically during the crash, requiring no action by the occupant. Because the driving public generally does not voluntarily use available belt restraint systems, American authorities have turned their attention to passive systems with the most interest being focused on the inflatable airbag. This system operates on a principle that sensors in the vehicle detect an unusually high rate of deceleration occasioned by the frontal collision. These sensors then cause a fabric bag to inflate rapidly with a loud noise. The inflated bag fills the space between the occupants of the front seat and the dashboard area of the car, cushioning them when they are thrown forward by the crash and distributing the energy dissipation over a wide area of their body. A similar bag can become inflated to cushion the forward movement of rear seat passengers. The American government and automobile industry are expediting research and development efforts on air-bag restraint systems.

Head restraints are also now required equipment in the interior of American vehicles. They can provide "whip-lash" protection for vehicle occupants during rear-end crashes. However, they often are inadequate because of improper adjustment and they compromise rear visibility for many drivers. For this reason the U.S. government is sponsoring a program to develop and demonstrate prototype deployable head restraint systems that automatically position themselves to prevent injury without significantly impairing driver vision during normal vehicle operation.

The National Highway Traffic Safety Administration of the United States in its *1970 Report on Activities* states:

The number of pedestrian fatalities in the U.S. during the past 3 years exceeded by 72% the total fatalities in all aviation, marine, railroad, and grade-crossing accidents combined, 29,000 compared with 16,900. In urban areas, where the bulk of our population lives, more than half of the traffic fatalities are pedestrians. Characteristics of the problem are described as follows:
All ages are affected, but over half of the pedestrian fatalities are in two age groups, those under 15 and those above 61. In the remaining age groups, alcohol is heavily involved.
Over half of the nonfatal pedestrian injuries occur to children under 15.
Ninety per cent of fatally injured pedestrians over 15 are reported as never having been licensed to drive; the pedestrian's unfamiliarity with the driver's task appears to be a very important factor.[5]

Because of this and similar data, officials of the U.S. and other nations are becoming more concerned with increasing pedestrian safety through such varied programs as improved pedestrian control methods and procedures, physically separating pedestrian and vehicular movements at high conflict locations (grade separations and malls), and expanded pedestrian education programs.

Postcrash phase countermeasures are primarily directed toward emergency signal generation and other communications, emergency transportation of the injured, emergency medical care, and debris removal so that the roadway can be restored to normal operation.

THE ROLE OF THE VEHICLE

As indicated in the previous discussion, vehicle design and operating characteristics play an important role in both the cause and severity of accidents. The U.S. National Highway Traffic Safety Administration sponsored 17 multidiscip-linary accident investigation teams which operated from university medical and/or research centers in various sections of the country. One of these teams had as its objective a determination of the extent to which vehicle defects, malfunctions, and vehicle subsystem maladjustment cause and contribute to crashes. The team's findings were based principally on its investigations of 50 crash cases. It was concluded that each case was probably caused by mechanical defects. The team report concluded that incompetent or incomplete servicing may be an important factor in such accidents. It was also reported that, in some cases, the owners themselves had either failed to recognize that their vehicles were in unsafe operating condition, or decided against having necessary repairs made.[6]

Another multidisciplinary team of investigators was assigned to investigate and determine the causes of motor vehicle crashes and occupant injury and to determine the relationship between vehicle defects, malfunctions, and subsystem maladjustments contributing to those crashes. This team questioned the hypothesis that primary vehicle defects are a significant cause of crashes. Seventy-six per cent of the experimental sample in this study had defects which caused or contributed to a crash. None of these defects were caused by a "lack of something necessary for completeness' (Webster's definition of defect) when the vehicle was new—a primary defect. The defect was a product of subsystem part degradation through prolonged use. It was also concluded that a special examination of "secondary defects" would help to reduce the incidence and severity of crashes. Brakes were the most significant subsystem that caused or contributed to the crashes investigated. A second group of "secondary defects" were categorized under the heading of "steering–suspension". Although this category did not cause crashes of the vehicles examined, it did contribute to 20·5% of those crashed. Included under this category were degraded power steering systems, idler-arm wear, worn bushings, etc. The team reported that tires were the only other subsystem that were rated as "causative" in a crash. They recommended that vehicle inspection procedures include adequate examination of the brakes, tires, and steering suspension systems.[7]

Research in crash victim protection has resulted in a number of rule-making actions by the U.S. government. These rules are vehicle safety standards. Their wide range of coverage is partially indicated by the following listing of recent changes, each of which increases the chances of crash survival:

Improved safety qualities of windshields (replacement of glazing materials which shatter upon impact into numerous small, sharp pieces that can seriously cut the occupants, by materials which tend to bulge out at the site of impact, but do not break).
Retarded flammability of interior materials.
Increased protection from roof collapse.
Improved child safety seating systems.
More stringent requirements for seats and seat belts in trucks and buses.
Improved safety qualities of side and rear windows.
Improved interior impact protection (padding).
Improved protection of driver from steering-assembly injury.
Increased rear underride protection.
Improved fuel system integrity (fire protection).
Improved side door strength.

The number of deaths and serious injuries that result from vehicle crashes can be reduced by controlling the forces to which vehicle occupants are exposed by improving the crashworthy quality of the vehicle. The crashworthy vehicle consists of two subsystems: (1) the occupant package inside the vehicle itself consisting of the restraint systems and the composition of the interior surfaces; and (2) the energy-managing structure of the vehicle. These two subsystems function together to provide a cushion for vehicle occupants during crashes. Distance is required to provide this cushion, and only a limited amount is available. U.S. crash survivability research is aimed primarily at finding ways to use the available distance efficiently, demonstrating the feasibility of greatly reduced crash injury. The occupant package currently is the most important element of the crashworthy vehicle.

The objective of the American vehicle structures program is to establish the feasibility of building automobiles that behave during collisions in a way that minimizes crash injury. This requires development and demonstration of structures that:

Prevent intrusion of the occupant compartment.
Control vehicle decelerations during a crash, so that, in conjunction with the occupant packaging system, crash forces and accelerations do not exceed human injury thresholds.
Avoid postcrash complications that precipitate new hazards (such as fire), or delay urgent remedial activities.

Another series of U.S. government-sponsored tests were directed toward determining the feasibility of designing a vehicle frame structure that would provide optimum deceleration rates for the full periphery of the vehicle under impaction with a solid object. The crashworthy qualities of an automobile structure greatly depend upon the way the kinetic energy of the vehicle is dissipated during various types of collisions. This capacity of the structure to dissipate energy is directly related to its force-deflection characteristics. Present automobile structures provide little structural resistance between the front bumper and engine. They also tend to exhibit linear force-deflection responses for increasing loads during low-velocity impacts. For these reasons changes in the structural response require some departures from present design practices. The front-end structure of a typical full-sized American production automobile has a distance of approximately 4 ft between the front bumper and the engine compartment fire wall. The engine, which is about 2 ft in length, behaves as a rigid mass during the collision, and the available acceptable collapse distance with current designs is restricted to about 2 ft, corresponding to an energy dissipation capacity for 40 m.p.h. impacts, if suitable structural modifications are made. Because of this, the research effort considered the problem at two levels: (1) modifying the structure so that the energy dissipation is primarily contained in the region in front of the engine and between the engine and the fire wall; and (2) modifying the engine mounting so that as it is pushed backward during the front-end collision, it would be deflected downward to prevent intrusion of the occupant compartment; or (3) removing the engine so that the entire 4 ft is available for collapse. Because less than 1 ft of useful collapse is available on the sides of the passenger compartment in existing vehicle configurations, side compartment modifications present a complex design problem. Structure modifications in three vital areas were considered, i.e. reinforcement of (1) door panels, (2) passenger compartment structure, and (3) the frame structures beneath the compartment.

The major result of this study is the clear demonstration that a significant improvement in the capability of the structure to dissipate energy in a controlled manner in front and side collisions is possible. All of the concepts, the three relating to front collisions as well as the one concerned with side impacts, show that for a given impact velocity a decrease in the vehicle collapse distance can be produced without significantly increasing the peak values for the compartment deceleration.[8]

In order to advance the chances of crash survival, several nations are now involved in the Experimental Safety Vehicle (ESV) program. The purpose of this program is to test new ideas of automotive

TABLE III Distribution of motorcycle accident involvements resulting in an injury or fatality by type of area in the United States, 1970

| Age of Motorcyclist | Rural areas | | | | | | Urban areas | | | | | |
| | Injury | | | Fatality | | | Injury | | | Fatality | | |
	Single veh.	Multi-veh.	Total	Single veh.	Multi-veh.	Total	Single veh.	Multi-veh.	Total	Single veh.	Multi-veh.	Total
Under 20	16·3	24·7	41·0	9·0	26·8	35·8	9·6	32·9	42·5	7·8	24·3	32·1
20–24	15·2	12·9	28·1	12·6	16·7	29·3	9·3	21·4	30·7	12·0	23·2	35·2
25–34	12·1	8·8	20·9	9·9	10·5	20·4	6·5	12·4	18·9	9·0	13·8	22·8
35–44	4·0	2·1	6·1	4·3	3·6	7·9	1·8	3·3	5·1	1·5	4·1	5·6
45–54	1·7	1·0	2·7	3·2	1·0	4·2	0·7	1·3	2·0	2·0	1·2	3·2
55–64	0·5	0·5	1·0	0·8	1·0	1·8	0·2	0·5	0·7	0·3	0·4	0·7
Over 64	0·1	0·1	0·2	0·3	0·3	0·6	†	0·2	0·2	0·2	0·2	0·4
Total	49·9	50·1	100·0	40·1	59·9	100·0	28·1	71·9	100·0	32·8	67·2	100·0

†Indicates at least one reported involvement but less than ·05 percent.
Source: *1970 Report on Activities Under the National Traffic and Motor Vehicle Safety Act* (U.S. Department of Transportation, National Highway Traffic Safety Administration, Vol. II, Washington, D.C.).

safety incorporated in a complete vehicle that will be designed, fabricated, and tested as a total system. The U.S. government is working with private contractors who are designing and producing prototypes in the 4000 lb vehicle weight category. All of the contractors are designing to conform to performance specifications which require that drivers and passengers must be protected from serious injury or death in a head-on barrier crash at 50 m.p.h., in a side impact with a rigid barrier at 15 m.p.h., and in a rollover at 70 m.p.h. West Germany and Japan are now developing ESVs in the 2000 lb class. These countries have formalized agreements with the United States for the exchange of technical data generated by the ESV program. Information exchange programs have also been signed between the United States and France, and the United Kingdom, Italy, and Sweden. Several European countries also have working cooperative agreements in this area.

Efforts have been made to reduce other potential dangers related to the motor vehicle, for example, in vehicle operating systems such as the vehicle controls and displays, vehicle lighting and visibility, handling and stability, and tires and wheels.

Although the United States had a decline in total traffic deaths in 1969–70, the motorcycle fatality toll increased from 1945 in 1969 to 2429 in 1970. The result was to raise the death rate per 10,000 motorcycles from 8·1 to 9·0. It is interesting to note that the rate is about 11% higher in the states which do not require motorcyclists to wear suitable helmets. The motorcycle death rate per

million miles of travel is nearly five times the rate for all motor vehicles. A U.S. research study was made of the effectiveness of using motorcycle headlights and taillights during the daytime as a crash-avoidance technique. The study investigated the effects of headlights and taillights on motorcycle accidents, motorcycle noticeability, and motorcycle electrical systems. The findings showed a significant decrease in daytime accidents in four states having daytime motorcycle headlight laws. In addition, the results indicated that daytime operation of motorcycle headlights increases notice-ability between 44% and 142%, depending upon the traffic situation. Most motorcycles appear to have electrical systems that are adequate for extended daytime use of the lighting system.[9]

As indicated in Table III, about 90% of all of motorcycle injuries and deaths in the U.S. involved persons under 35 years of age.

THE ROADWAY AND ITS ENVIRONMENT

No matter how well educated a driver may be, and no matter how well designed and maintained his vehicle is, his relative safety still will be greatly influenced by the physical features of the roadway environment in which he operates, including the travel-way, the road-side development, and such variables as weather and the presence of other traffic units. A detailed explanation of the roadway variable is beyond the scope of this paper. Through the years, numerous advancements have been made to improve the design and operating characteristics of the roadway including:

Development of geometric design standards for various functional classes of roads.
Construction of increased mileage of multilane controlled access roadways with no crossings at grade (freeways, autobahns, etc.).
Introduction of spacious medians in rural areas and median barriers in urban areas to physically separate opposing traffic movements.
Improved design construction of guard rails and other vehicle barriers.
Improved road surfaces.
Fewer distracting influences, such as billboards and other advertising signs, merchandise displays, and people.
Removal of hazards from close proximity to the roadway and the use of "breakaway" designs for those obstacles which must remain close to the road.

For a more detailed discussion of this subject, the reader is referred to publications of governmental agencies and the Road Research Laboratory (U.K.), Highway Research Board (U.S.), and similar organizations.

GENERAL COMMENTS

The concurrent development of improved accident reporting procedures, multidisciplinary crash investigation study teams, and electronic data processing procedures has and will continue to enable man to better understand the complex phenomena of highway traffic safety. Two excellent state-of-the-art papers on motor vehicle accident studies trace the history of motor vehicle accidents through various eras of concentration to trends for the 1970's.[10,11]

One example of mass data analysis was recently reported in the United States. The National Accident Summary File was analyzed to determine the one-way relationships between nine independent variables and the severity rates. Thus, for the crashes in this subset, various comparisons of crash severity are possible.

The Accident Type and Location statistics indicate that single-vehicle and rural crashes have higher severity regarding both fatalities and injuries. The Collision Type figures show that pedestrian crashes have the highest fatality severity (although not the highest injury rate), while angle and rear-end crashes have the lowest fatality rate. Crashes identified as running off the road and striking fixed objects have the highest injury severity. The lower severity rear-end and angle crashes account for approximately 75% of the crashes in the file. Dry roads have the highest severity crashes while snowy and icy roads have the lowest, no doubt due to

differences in velocity. Crashes which occur after dark have higher severity. Similarly, weekend crashes have a greater severity than do weekday crashes.[12]

It is not only important that each nation become more concerned with improved transportation safety, but it is essential that there be greatly increased international cooperation in order to minimize the duplication of effort and to provide for the free interchange of information among all peoples. The motor vehicle has become the main transportation mode in the world and it will continue to be such for many years to come. In the interest of conserving human and natural resources throughout the world, all nations should be directly concerned and involved in traffic safety because of their responsibility to their own citizens and citizens of other countries.

REFERENCES

1. G. M. Bastarache, "Drunk driving: the European view" *Highway User*, November 8–11. Highway Users Federation, Washington, D.C. (1970).
2. S. Schwimmer and R. A. Wolf, "Leading causes of injury in automobile accidents" *ACIR Report*. Cornell University, New York (1962).
3. *Ibid.*
4. *Safety Belts—Their Fitting and Use*. Institute for Road Safety Research SWOV, Voorburg, The Netherlands (1970).
5. *1970 Report on Activities Under the Highway Safety Act of 1966*. U.S. Department of Transportation, National Highway Traffic Safety Administration, Washington, D.C. Vol. II, p. 20.
6. *Relationship between Vehicle Defects and Vehicle Crashes: Final Report*. Stanford Research Institute, Menlo Park, California (1970).
7. *Multidisciplinary Investigations to Determine Relationship between Vehicle Defects, Failures, and Vehicle Crashes: Final Report*. Baylor College of Medicine, Department of Psychiatry, Houston, Texas (1971).
8. *Basic Research in Automobile Crashworthiness: Summary Report*. Cornell Aeronautical Laboratory, Inc., Buffalo, New York (1970).
9. *Daytime Motorcycle Headlight and Taillight Operating*. The Franklin Institute Research Laboratories, Philadelphia, Pennsylvania (1970).
10. W. G. Eames, N. L. Scott and J. C. Fell, "Motor vehicle accident studies" *1970 International Automobile Society Conference Compendium*. Society of Automotive Engineers, Inc., New York (1970).
11. K. J. B. Teesdale, "Vehicle accident studies: state-of-the-art review (Non-U.S.A.)" *1970 International Automobile Society Conference Compendium*. Society of Automotive Engineers, Inc., New York (1970).
12. W. L. Carlson, "AID analysis of national accident summary data" *HIT LAB Reports*. The University of Michigan, Highway Safety Research Institute, Ann Arbor, Michigan (Oct. 1971).

TRAFFIC ACCIDENTS—A MODERN EPIDEMIC

MURRAY MACKAY

Department of Transportation and Environmental Planning,
University of Birmingham, Birmingham 15, U.K.

Accidental trauma in general and traffic accidents in particular are discussed briefly within the framework of the host, agent, environment complex of conventional epidemiology. The rise of accidents is contrasted to the fall in infectious disease over the years, and then some of the basic characteristics of traffic accidents are reviewed. It is suggested that programmes which aim at behavioural modification are unlikely to produce startling improvements in the short run. Similarly the economic restrictions on environmental changes would seem to inhibit radical benefits from being achieved in the immediate future. The possibilities of injury reduction rather than crash avoidance are then discussed, and improvements in this area are suggested as having the greatest effect within the next few years. The emerging science of accident research is mentioned as an appropriate and distinctive field of study which should be encouraged.

INTRODUCTION

Accidental death and injury, unlike many conventional diseases, has not yet come under the objective scrutiny of the scientific world. The very term "accident" implies some special non-rational event, in which the normal causal relationships are suspended, or are somehow intrinsically different from those which lead to other forms of mortality or morbidity. Involvement in an accident is still often explained away by laymen in terms of "luck", an "act of God", or, to those who have survived a military collision, "the Jesus Factor".

If accidents are indeed fundamentally different from other natural phenomena, then their eradication can safely be left in the hands of the philosopher and the clergyman. If on the other hand accidents are subject to the normal causal relationships which govern all our external activities, then the methods of epidemiology, the evaluation of the probabilities of accident involvement associated with various levels of exposure and risk, can appropriately be applied to the study of accident phenomena.

Although such an approach, together with its implications for government and industry, is still not yet universally accepted, there is enough information available to describe in general terms the nature of this traumatic epidemic which stems in large part from the motor vehicle.

Having survived birth, for over half their lives, most people in the western world are more likely to die from accidents than from any other cause. In the world as a whole, only cancer and cardiovascular disease cause more deaths than accidents.[1]

A useful but sometimes misleading concept in establishing priorities is to consider the number of life-years lost from various causes. In comparison to the diseases of degeneration, accidents, because they happen predominantly to young adults, become more important than straight numbers of deaths, and for most of the developed countries rank first, above all other sources of mortality. Such comparisons can be misleading, but if the economic values implicit in such analyses are thought to be appropriate in establishing priorities for spending public funds, then the conclusions are inescapable.

This high ranking of accidents as a cause of death has not resulted however from a large increase in their incidence. Rather, the numbers of accidents has remained relatively static over the last 70 years (although varying in composition) in most western countries, and the more conventional modes of death, particularly infectious diseases, have declined. Vaccines, antibiotics, surgical advances, pure water supplies, improved sanitation and general advances in diet and living standards have virtually eliminated deaths from typhus, smallpox, diphtheria, influenza, cholera and other diseases which were part of the everyday world of the last century.

Accidents take many forms, and of the total those associated with motor vehicles constitute approximately one-third in the western world. The other large single category of accidents consists of falls (22%); the remainder resulting from fire, drownings, other transport modes, poisoning, firearms, etc.[2] It is worth noting that conventionally we define an injury-producing traffic accident as that event which results from a conflict between road vehicles, between a vehicle and a pedestrian or a single vehicle accident in which an occupant is injured. This definition is vehicle based, but the actual mechanism of injury is not categorized. It is of more than passing interest to the pedantic to note that in fact a large number of falls occur in the road environment. These are events in which a pedestrian is injured, and they are influenced by factors which are within the normal compass of the traffic engineer—the design of pedestrian facilities, kerbs, ramps, steps, access routes to public buildings and loading arrangements for P.S.V.'s. In many ways it would be logical to include these "single vehicle" pedestrian only accidents with other traffic accidents if one wishes to describe the total epidemiology of trauma in the traffic environment, including both wheeled and foot traffic. If such a definition were adopted then the numbers of traffic casualties would increase considerably.

Thus traffic accidents, even conventionally defined, as a cause of death and injury pervade our present society. Using somewhat arbitrary assumptions, the cost in Britain to the community of traffic accidents is at least £300,000,000 per annum.[3] For every death there are 13 serious injuries (persons requiring in-patient treatment in a hospital) and 38 minor injuries (persons requiring out-patient treatment in a casualty clinic).

A prediction has been made of the likely numbers of road casualties in Great Britain for the year 2000 A.D. (see Table I). Corrections from the present ratios of fatal : serious : slight have been made in view of the likely effects of future developments in the prevention and reduction of fatal and serious injury by improved vehicle design. Also

TABLE I

Traffic accidents in Britain in 1968 and in 2000 A.D.

	1968	2000
Fatal and serious casualties	95,373	200,000
Total casualties	349,208	1,000,000
Total injury producing accidents	264,200	600,000
Total number of cars	8,750,000	24,000,000
Total number of vehicles	13,903,000	33,000,000
Vehicles per head of population	0·26	0·53
Cars per head of population	0·16	0·39
Population	54,000,000	62,000,000

the increasing density of traffic which comes with higher levels of ownership of itself will effect the outcome, as will the projected road programme of the nation, but even this fairly optimistic view suggests that the treatment and rehabilitation of traffic accident victims will become one of the dominant activities of the medical profession. If such a situation prevails then on average one driver in three can expect to be involved in a serious or fatal injury accident during his driving life.

EPIDEMIOLOGICAL CONSIDERATIONS

A somewhat loose parallel is often drawn between the host, agent, environment divisions of classical epidemiology, and the road user, the vehicle and the highway of traffic accidents. In some ways, however, a more fruitful conception of an accidental event has been put forward by Haddon.[4] That author has pointed out that accidents (of all types) result from abnormal energy exchanges, be they chemical, ionizing or electromagnetic radiation, or mechanical, as in the case of traffic accidents. This energy exchange corresponds to the agent of infectious and other diseases.

One may then consider a road accident in the form shown in Figure 1. Such a model has the advantage of taking the emphasis away from the purely causal aspects or pre-crash phase, and allows alternative strategies for the amelioration of the event to be considered. In particular recent

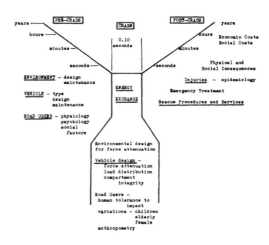

FIGURE 1   The anatomy of a road accident.

research has suggested that major benefits can be realised by modifying the crash phase conditions, and there is now a world wide governmental and industrial effort being made to realise some of the potential benefits of improvements in this area. Just how far the concept of protection against harmful energy exchange can be taken is under debate at the present time, and pinpoints one of the aims of future accident research.

For example, for an increase in price to the consumer of approximately 30% on present day car costs, it is possible to save at least half of all serious and fatal car occupant casualties.[5] On almost any cost benefit criteria, such an equation appears to be a favourable proposition. However, to save 80% of serious and fatal car occupant casualties might well double the present price of a car. Thus, there is some point beyond which improvements in the crash phase of accidents become marginal and alternative strategies have to be explored. The present state of knowledge is not yet sufficient to allow such alternatives to be defined.

THE HOST

The characteristics of those who are involved in traffic accidents can be described in general terms, as can the surrounding circumstances. In Britain car occupants and pedestrians are killed or injured in approximately equal numbers—some 40% of the total for each category, the remainder being riders of two wheeled machines, mainly motorcyclists. Car occupants are predominantly young males—the 18-25 year age group have an involvement rate some four times the mean for all ages. By contrast pedestrians represent the young and the old, with one third of serious pedestrian casualties being under ten years of age and a further third being over 55 years. Males predominate over female pedestrians by a factor of 2.

Alcohol is present in fatally injured car occupants with significant frequency—approximately 30% of such occupants are grossly impaired by alcohol, although this figure may have been modified in recent years due to the Road Safety Act of 1967. Alcohol is also thought to be important in adult pedestrian fatalities but there is no research data in this country to evaluate this aspect.

With the exception of alcohol, there are no stable or predictable road user characteristics which can be identified with a large proportion of traffic accidents. Behavioural research has shown that certain physical, psychological and social characteristics lead certain groups of the population to high traffic accident involvement. But the main indication from behavioural research is that there is no single characteristic the eradication of which would solve the accident problem. For example, young, extrovert, unmarried manual workers exhibit a life-style which includes above average traffic accident involvement. The elimination of such a group from the highways, however, might well have only a marginal influence (less than 10%) on the overall accident picture, even assuming that such a group could be identified practically. The message of present day behavioural research is that there is no solution of such power, which would also be politically and socially acceptable, which would radically reduce the problem.

In the long term, it would seem reasonable to suggest that behavioural modification must develop through better selection and training procedures for drivers, improved vision and medical tests, greater control of driving hours, regular retesting and restrictions on night driving by the elderly, and many other programmes which identify conditions of high risk. But to the extent that driving is an expression of an individual's life-style, it would seem that risk taking and the occasional act of incompetence are fundamental, and we should not therefore expect too much from programmes which aim to modify behaviour, assuming that the next generation or two are fundamentally similar to ourselves.

ENVIRONMENTAL ASPECTS

One of the main factors in the control of infectious disease was the control of the environment, through water purification, sewage disposal, elimination of favourable breeding grounds for mosquitoes and many other changes which influenced the habitat of the agents of diseases. So too with accidents, the environment exerts a controlling influence over their occurrence. Indeed, in an academic sense one may say that any environment which allows two vehicles to conflict with each other is deficient. And some purists remark that the railways, not noted for their enthusiastic introduction of safety measures in the past, gave up the idea of trying to get two vehicles travelling in opposite directions to share the same track in the very early days of rail transport.

In practical terms however, we have an invest-ment in a road network which began in Roman days and has continued through the development of Medieval towns, through the upsurge of the Victorian industrial period to the present. This environmental legacy means that even the most grandiose road programme is only making a very small difference to the overall network. Even allow-ing for a traffic distribution which shows that 80% of the traffic occurs on 20% of the roads, the rate of change of the driving environment is slow.

Because of this historical background there are a host of environmental measures which are known to reduce the occurrence of accidents, and which are not yet applied simply because funds are not available. The introduction of street lighting, improved junction layout, the use of high skid coefficient road surfaces, median barriers, pedestrian phased traffic signals and the like represent short term improvements. Grade separation, pedestrian and vehicle segregation, and limited access roads represent the long term measures, with automatic, computer-run, driving offering a final, utopian solution. Technically, automatic control of vehicles is already perfectly possible; the introduction of such a system however presents formidable political and economic problems which suggest that it is unlikely to affect a significant amount of our road network before 2000 A.D.

THE AGENT

We do not know when or where the next road accident will take place, nor who will be involved in it. But we can be certain that death or injury will result because a part of the human frame has received an amount of mechanical energy in excess of a tolerable level. Hence, the importance of packaging the road user so that he receives only tolerable amounts of energy.

In a collision there are two parts, first, the vehicles interact and lose their kinetic energy mainly by plastic deformation of their structures. This is the first collision. After this part of the crash is largely over, then the occupant strikes the interior parts of the vehicle, the second collision. The forces operating during this second collision cause the actual trauma. These forces are localised on certain parts of the body, the head, chest or lower limbs, and are of short duration and of extremely high magnitude. It has been established from the work of John Stapp[6] and others that the primary forces

acting in the majority of collisions are survivable. Hence, the aim of good crashworthiness design is to restrain the occupant and attenuate his impacts with interior structures.

Unfortunately the real world of accidents shows a large range of collision configurations, impact speeds, striking objects and complex dynamic circumstances. What may be appropriate for one type of accident, provides no benefit or even a liability in another type. Many of the simpler crash performance criteria in car design are now being met in current model cars as far as the occupants are concerned, but until better descriptions of the frequency and severity of specified conditions in actual accidents are provided by field accident research, then developments will remain piecemeal and far from optimal.

This may be illustrated by considering the differ-ing requirements of occupant and pedestrian pro-tection. In Britain there are some 28,000 fatal and serious pedestrian casualties each year, and some 38,000 fatal and serious car occupant casualties. Almost all pedestrians, when in conflict with a car, are hit by the front structure. Car occupant acci-dents on the other hand, represent a range of collision types of which only some 55% are frontal.[7] Therefore it would appear that the front structure of a car should be thought of firstly as a structure with which to hit a pedestrian, as it does that more frequently than attenuate the collision forces for the occupants. (Particularly as only some 12% of front seat occupants are currently restrained by seat belts, and thus utilise the benefits of collapsible front structures.)

Until a great deal more information is available from traffic environments throughout the world, the optimum resolution of such design conflicts will remain unsolved. It is to be hoped, however, that one of the effects of a very active legislature in the United States in the field of regulation of the crash performance of vehicles, which has cata-lytically led to governments elsewhere instituting similar programmes, will result in thorough research so that these programmes are scientifically sound and appropriate to the countries concerned.

THE EMERGING SCIENCE OF ACCIDENT RESEARCH

Severy[8] has noted that it is dangerous to be alive, but the alternative makes that hazard endurable for most of us. Now that medicine almost guarantees survival for a biblical lifespan, the hazards of accidental trauma have become of major concern to the advanced western countries. As yet however, no established scientific precedents exist to show how this modern traumatic epidemic may be curtailed. The problem requires the development of a new subject—a scientific approach to accidental phenomena of all kinds, the most pressing at the moment being road traffic accidents.

This class of accident however is only one of many types, and the objective study of accidents should encompass all modes of travel, as well as home, industrial and recreational activities. The overall aims are to define levels of exposure of various groups, establish risks for given exposures, describe causal relationships, examine mechanisms of injury, describe the consequent trauma, suggest remedial modifications and monitor the effects (beneficial or otherwise) of any changes which occur.

It is to be hoped that objective evaluation and analysis of accidental phenomena will replace the conventional ethic which still in large part centres on causes, and pins much of its hope for success on behavioural change brought about by exhortation. Such programmes, when studied objectively often show no benefits, and indeed can do harm by diverting attention away from the real aetiology.

Quality of life issues are now receiving attention on a national scale. It would seem that the preservation of life, the reduction of morbidity and disability as a result of accidental trauma are absolutely fundamental to the quality of life, as we now aim to plan it. Therefore it is to be hoped the emerging science of accident research will shortly develop, first the tools and then the solutions for that most modern of epidemics, the road accident.

REFERENCES

1. World Health Organisation, *Annual epidemiological and vital statistics*, Geneva (1969).
2. R. A. McFarland, "*Health and safety in transportation*" *Public Health Report* 73(8), 663–680 (1968).
3. R. F. R. Dawson, *The Current Cost of Road Accidents in Great Britain*. Department of Environment, Road Research Laboratory Report LR 396 (1971).
4. W. Haddon Jr., E. A. Suchman and D. Klein, *Accident Research*, p. 537. Harper & Row, New York (1964).
5. G. M. Mackay, "Safer cars by 1977" *New Scientist* 210–214 (Jan. 27, 1972).
6. J. P. Stapp, *Principles of Automotive Crash Protection*, p. A1.3–20. Proc. Conf. Road Safety, Brussels, Belgium (1968).
7. G. M. Mackay, "Some features of traffic accidents" *Brit. Med. J.* 4, 799–801 (Dec. 57, 1969).
8. D. W. Severy, H. M. Brink and D. M. Blaisdell, *Small Vehicle versus Larger Vehicle Collisions*, p. 2. S.A.E. Paper No. 710861, New York (No. 1971).

THE DESIGN OF HYBRID CUSHION CARS

DAVID FOSTER

White House, Sunninghill Road, Windlesham, Surrey, U.K.

The new American car safety legislation, Standard 208, is "performance-based" in specifying a car crash speed up to which the occupants will not be injured but no indication is given *how* to do it. The two leading contenders for the actualization are the inflatable air-bag and The Hybrid Cushion Car invented by the author.

The principle of the Hybrid Cushion Car is to combine safety harnesses with deep internal paddings. The design is fully optimized in a specific mathematical formula which includes the Newtonian laws for retardation and also the established safe medical values as to non-injurious impact levels.

Much of the design has a common philosophy with the techniques used by NASA to bring spacemen down safely at splashdown. The indications are that the majority of the worlds car industry will use the Hybrid Cushion Car design as the means to actualize Standard 208 and the equivalent legislation being prepared in other countries.

In October 1970 a quite new type of car safety standard was passed in Washington and is mandatory on all new cars by 1975. Previous safety standards had specified the *means* to accomplish safety such as by wearing safety belts, by the provision of anti-burst door locks, by the use of collapsible steering columns and so forth. This type of legislation is "engineering-design based" and it tells the car-maker how to design his car to make it safer. However, the facts of the statistics shewed that car deaths and injuries were not falling in spite of these measures and in the United States now run at about 5,000,000 injuries a year and thus an entirely new form of safety legislation was introduced which simply stated that in a specified car crash the occupants would *not be injured at all*. The new Standard 208 requires that occupants shall not be injured in frontal crashes at up to 30 m.p.h. and such legislation is called "performance based". No recommendations are given by the U.S. Government as to how to achieve the safety performance and the legislation of the critical speed is related to a consensus of opinion by road safety scientists that "they are sure it can be achieved". Naturally the higher the legislated safe speed then the safer the car over the total injury and crash spectrum.

THE TWO CONTENDERS TO MEET STANDARD 208

In the event two distinct safer car designs have emerged to meet the requirements of Standard 208 and they are:

(a) The inflatable air-bag, probably to be used with safety belts in addition.
(b) The Hybrid Cushion Car which consists of using safety belts in a highly padded car interior and mathematically optimized to a Formula MSL-100. The author's name is associated with the origin of this design.

THE INFLATABLE AIR-BAG

Perhaps it would be appropriate to say a few words about the inflatable air-bag technique. This consists of a small cylinder of gas compressed at 3500 psi and connected to a folded plastic bag in the dash area or in the centre of the steering wheel. At an impact greater than a predetermined level the gas is released by sensor triggering and expands in front of the occupants to create a cushioning balloon and then collapses. It has been criticized of this system that:

(a) people would be apprehensive of having with them in the car what is virtually a small bomb and people can be hurt if they are not seated in their normal position.
(b) the noise generated is some 160 dB, some 20 dB over the threshold of pain and thus rather dangerous to the hearing.
(c) when the air-bag goes off then driving control is lost and thus if it goes off accidentally a dangerous situation could be created.
(d) the air-bag is only designed against forward impact and can make no contribution to other

impacts such as sideswipes, rear collision or
rollovers.
(e) the problem of maintenance is considerable and
it has been suggested that the device would
need to be maintained to "better than Aero-
Space standards". One wonders where the
skilled staff are to come from.

Perhaps the most damaging aspect is that, at the
time I write, no volunteers have been found willing
to test the system. From a control system aspect it
would appear to be doubtful if "the bomb of a car
crash" can be accurately neutralized by a counter-
bomb.

THE THREE COLLISIONS IN A CAR CRASH AND THEIR SAFE MEDICAL CRITERIA

There are three collisions in a car crash involving
occupants and they are as follows:

The First Collision

This is when the car strikes some object such as
another vehicle or a stationary object and the only
injury is to the car itself and it thus has no relevant
injury criteria *per se*.

The Second Collision

With the car halting or halted, the inertia of the
occupants is unaffected and they continue to move
forwards at the pre-crash speed in the absence of
slowing restraints. Thus the occupants crash into
the inside surfaces of their vehicle and this is the
Second Collision. Since the car is harder than the
human being the occupant will tend to be injured
rather than the car. Note that the injurious forces
are created by the inertia of the occupants and will
be directly proportional to their weights.

The author has discussed the question of injury
criteria in second collision with the leading experts
on the matter who advise that the human body
should not be subjected to *pressures* of greater than
30 psi with the critical zone being the soft abdomen
which does not have bone protection so that pressures here can rupture internal organs such as the
liver, stomach, spleen and intestines. It is also
significant that this is the zone where occupants
often wear their lap belts and we shall later see that
this possibility has to be accounted for.

The Third Collision

When the occupant is stopped in Second Collision
his internal organs are still somewhat free to keep
moving forwards under their own individual inertia
forces and there thus can be a Third Collision
between such organs and the interior walls of the
body. For example the heart can bang against the
inside of the chest wall and in that case damage is
likely to be due to heart vessel rupture since such
vessels are trying to hold the heart back like mini-
safety belts. Also typical of Third Collision injury
is that due to the jelly-like brain surging forwards
under inertia forces and creating concussion and
other damage.

The medical criteria for safety at Third Collision
is well established by workers such as Col. John
Stapp who took sled ride decelerations for the U.S.
Air Force in California and it is now generally
agreed that the onset of such injury takes place at
$45G$ and that $40G$ is a reasonable maximum safe
design figure.

The G-value is a measure of the sharpness of an
impact and, for example, if one jumps off a chair
onto the floor the G-value will be about $3G$.†

Thus the two medical criteria to ensure an occupant is safe in a car crash is that he shall not be
subjected to pressures greater than 30 psi nor G-
values greater than $40G$.

THE SAFE-STOPPING SYSTEM CONCEPT

In 1969 the author was kindly afforded facilities to
study the techniques used by NASA in bringing
astronauts down safely at splashdown which is
about a 30 m.p.h. impact into the sea and with the
latter somewhat "solid" especially in calm conditions. No-one has ever been hurt in such a landing.

NASA use a two-stage hybrid stopping system
in which the astronauts lay on padded couches
which themselves are suspended in a shock-
absorbing fashion, for example by nylon ropes.
There appear to be two distinct reasons for choosing
a hybrid system for human safe stopping as follows:
(a) One needs a primary retardation system in
order to do the main phase of stopping and this
is related to keeping G-values low. This relates
to the $40G$ criteria.
(b) But in addition the astronauts need to have
their bodies supported over a maximum area

---

† The terminal jump velocity is about 10 ft/sec and the
stopping distance typically 0·5 ft due to knee bending. The
equation is: $V^2 = 64GS$ where V is terminal velocity, and
S is stopping distance.

to reduce stopping *pressures* related to a given total stopping force. This relates to the 30 psi criteria.

In addition there are two side-advantages. In the first place any series-arranged hybrid system is a backup arrangement so that partial failure of one system may be partially compensated by the other system. But in addition if the crash is more than the designed level (say the space-craft came down on a rock instead of water) then the primary system would probably be over-extended and thus the secondary system takes the last crunch at a higher level than its designed operation but still safe if a margin has been built into the system.

Overall the author was considerably impressed by this hybrid approach which incorporated a "belt and braces" philosophy but it had the merit of possibly being able to be mathematically optimised as a true "system".

THE HYBRID CUSHION CAR

Taking advantage of the NASA philosophy, the author designed a safe stopping system (Figure 1)

FIGURE 1 Simple Cushion Car. The Hybrid Cushion Car has automatic or legislated use of safety harnesses in addition.

based on the following two hybrid components:
(a) *Car padding.* The car is internally padded so that occupant Second Collision impacts cannot exceed 40G at an impact velocity of 20 m.p.h. This means that the occupant is safe without other means of safety protection at a Primary

Collision of 20 m.p.h. It can be shown that this calls for about 5 in. of padding having a crushing strength of about 20 psi, i.e. having a safety margin of softness less than the 30 psi value from the medical criteria. This means that on human impact it is the padding which is "injured" providing it is not impacted at more than 20 m.p.h.

(b) *Safety harnesses.* The effect of a safety harness in a car crash is to restrain the occupant from hitting his own car inside in a Second Collision. But in the Hybrid Cushion Car they are worn for a somewhat different purpose. In a crash the safety belts restrain and *slow down* the speed of the occupant and if they can slow him down to 20 m.p.h. before he impacts the paddings then he will be quite safe for the reasons given in (a) above.

Thus in a Hybrid Cushion Car one may permit the occupant to have a Second Collision but *safely* into paddings at such a speed that the paddings are totally effective, i.e. at less than 20 m.p.h.

Obviously the system is thus safe for impact Primary Collision speeds well in excess of 20 m.p.h. and it will later be shown that the maximum safe speed can be exactly calculated.

STOPPING SAFETY BELT INJURIES

The statistics show that occupants wearing safety belts suffer, on spectrum average, only 50% of the injuries to a non-belted occupant and this was one reason why the author insists on their retention. But there is still that residual 50% of injuries caused to occupants wearing safety belts and these are mainly due to:
(1) Safety harnesses do not prevent the legs flailing forward and the Road Research Laboratory have commented "Leg injuries are surprisingly high to belted occupants". In a Hybrid Cushion Car such injuries are eliminated by the presence of the paddings.
(2) At higher speed crashes the restraint of safety belts can cause its own injuries. Existing lap and shoulder harnesses have a restraining area of about 60 in.^{2} Thus in a typical crash at 40 m.p.h. with the occupant coming to rest over a stopping distance of 3 ft, this corresponds to an average retardation of 18G and if the occupant weighed 160 lb then the average restraining force would be 2870 lb. Using a 2 in. wide

harness of 60 total in² then the average restraint pressure is 48 psi. Now this is well above the permissible medical limit of 30 psi and thus injuries are likely to occur. In such a case to avoid safety harness injuries the occupants retardation should not have exceeded $11G$ or a restraint force of 1800 lb.

Now in the case above there is a conflict of problems in that the occupant will suffer safety belt injuries but no second collision injuries. But if we arrange to have his safety belt extend at the safe value of $11G$ or 1800 lb then he will have no safety belt injuries but will suffer second collision injuries.

Let us assume we have arranged the occupant to be restrained at $11G$ whilst the car is retarding at $18G$. This means that the occupant is still moving after the car is stopped. But in the Hybrid Cushion Car he is going to hit paddings and thus the safety problem is whether he will hit them at less that 20 m.p.h., the safe value.

Applying a little mathematics it can be shewn that in the case cited he would hit the paddings at 17 m.p.h. and *thus would be quite safe*.

Thus in a Hybrid Cushion Car the safety belts are arranged to extend at a load in the region of 1800 lb to prevent excessive and injurious bodily pressures and the safe stopping is thus *shared* by safety belts and paddings in an exactly equivalent way to the hybrid systems used in space capsules at splashdown.

THE MATHEMATICAL OPTIMIZATION OF HYBRID CUSHION CARS

The combination of safety belt restraint and paddings can be optimized by writing a safe stopping equation which ensures that at no time has the occupant suffered pressures greater than 30 psi nor retardations greater than $40G$.

The Newtonian stopping equation for a two phase stopping is:

$$V^2 = 64(G_1 S_1 + G_2 S_2)$$

where V is collision speed in ft/sec

G_1 is retardation due to safety belts over a distance S_1 in feet

G_2 is final retardation due to padding over a padding thickness S_2.

The first item on the right-hand side of the above equation $G_1 S_1$ for safety belt slowing down has to be converted to relate to safe stopping pressures rather than G values via the intermediate formula:

$$\text{Stopping Force} = G_1 W = PA$$
$$\text{or } G_1 = PA/W$$

where P is the safe medical pressure of 30 psi

A is the safety harness restraint area such as 60 sq. in²

W is occupant weight in lbs.

In addition we have to evaluate the safety harness stopping distance S_1 which will be the crushability (X) of the car at the crash speed V (as determined experimentally) plus the distance (L) separating the occupant from the paddings when he is normally seated.

Thus the equation converts by substitution into:

$$V^2 = 64[PA/W (X + L) + G_2 S_2] \ldots \text{Safe Stopping MSL-100 Equation.}$$

Assuming that P is the safe medical value at 30 psi and the G_2 is the safe impact shock at $40G$ and that the occupant weight is 160 lb, the following are typical solutions for the thickness of required paddings in inches.

TABLE I
Padding thicknesses and crash speeds

Crash speed (m.p.h.)	Belt width (in.)	Car crushability in feet	Padding thickness
40	2	2·0	6·1
		2·5	4·4
		3·0	2·7
50	3	2·5	6·5
		3·0	3·9
60	4	3·5	6·0

Thus the practical range of operation of the Hybrid Cushion Car for *absolute safety* is in the range 40–60 m.p.h. and the higher speeds call for safety harnesses of greater restraint area and increased car crushability. It is forecast that such cars would be at least ten times as safe as existing cars without safety harnesses and five times as safe as existing cars using safety harnesses.

In this article the author has essentially dealt with the basic principles of designing safe stopping systems but he considers it not difficult to actualize them in practice and many car makers have such designs at a late stage of development.

Nor has the author dealt with the geometry of paddings although Figure 1 shows that the design consists essentially of a padded lap bench of such shape that on impact the occupant cannot reach the windscreen with his head. All "hardware" is banished to the periphery out of reach or is soft engineered. Notably the steering wheel is of a sprung-rim nature and lightly descends into an annulus in the paddings at low impact force of about 20 lb.

In the paper the author has only dealt with frontal crashes but the same principles apply to crashes from all directions as to the fundamental mathematics involved for absolute safety.

PART THREE

Legislation

THE IMPACT OF AUTOMOTIVE TRANSPORTATION ON THE ENVIRONMENT AND LEGISLATIVE MEASURES FOR ITS CONTROL
United States Experience

J. C. D. BLAINE

Dept. of Economics, The University of North Carolina at Chapel Hill, North Carolina 27514, *U.S.A.*

The central theme of this article relates to the legislative steps taken by the Government of the United States and its promulgation and enforcement of environmental quality standards in response to the nationwide concern with respect to the detrimental impact of automotive transportation upon the environment. Three aspects of this degrading impact are covered: pollution of the atmosphere by emissions from motor vehicles; scarring of the natural beauty of the landscape contiguous to highways, and the emission of physiologically harmful noise by automotive vehicles moving upon highways and streets, especially heavy duty trucks and tractor-trailer units. The concluding sections direct attention to the need for cooperative action on the part of the Federal, State and local governments, automotive and oil industries and an aroused populace in the formulation and the rigid enforcement of corrective environmental policies and practices, pertaining to automotive transportation.

NATIONAL ENVIRONMENTAL POLICY

The abusive manner in which this Nation has exploited its natural and bountiful environment is having telling effects upon the health and welfare of its people. So great is the impact that it is questionable whether the environment can continue to support the increasing industrialization and urbanization, unless it foregoes its ruthless and shortrun destructive attitude toward its environment. In response, there has arisen throughout the Nation demand for action to maintain and preserve an environment conductive to the health and welfare of the public.

The environment has been given low priority in the American economy, which is highly dependent upon energy derived basically from the incomplete combustion of fossil fuels. It is further characterized by rapid urbanization and a consumption pattern based in general upon a throwaway psychology. As a consequence, huge and increasing quantities of residue are thrusted into the environment daily. This degradation of the environment will continue unless stricter and more effective measures of control are enforced with the support of Federal, State and local governments, industry and a concerned public at large.

The need for a more integrated approach to the Nation's environmental problem was recognized by the passage of the *National Environmental Act of 1969.*[1] It sets forth the Congressional declaration of the Nation's environmental policy to create and maintain conditions under which man and nature can exist in productive harmony and fulfill the social, economic and other requirements of future generations. It also provides for the establishment of a Council of Environmental Quality, in the Executive Office of the President to assist in carrying out that policy. In addition, it is to assist and advise in the preparation of the Environmental Quality Report to be submitted annually to Congress; review and appraise programs; develop and recommend national policies, and conduct investigations relating to the environment. The detrimental effects of automotive transportation on the environment are part of its overall concern.

MOTOR VEHICLES AS A SOURCE OF AUTOMOTIVE CONTAMINATION

Transportation is a major source of air contaminants, especially carbon monoxide and hydro-

TABLE I

Nationwide emissions of contaminants, by sources, 1968

Sources	Carbon monoxide (Amt)	(%)	Particu- lates (Amt)	(%)	Sulfur oxides (Amt)	(%)	Hydro- carbons (Amt)	(%)	Nitrogen oxides (Amt)	(%)
Transportation	63·8	63·7	1·2	4·3	0·8	2·4	16·6	51·9	8·1	39·3
Fuel combustion in stationary sources	1·9	1·9	8·9	31·4	24·4	73·5	0·7	2·2	10·0	48·5
Industrial processes	9·7	9·7	7·5	26·5	7·3	22·0	4·6	14·4	0·2	1·0
Solid waste disposal	7·8	7·8	1·1	3·9	0·1	0·3	1·6	5·0	0·6	2·9
Miscellaneous	16·9	16·9	9·6	33·9	0·6	1·8	8·5	26·5	1·7	8·3
Total	100·1	100·0	28·3	100·0	33·2	100·0	32·0	100·0	20·6	100·0

(millions of tons)

Source: *Nationwide Inventory of Air Pollutant Emissions, 1968*, United States Department of Health, Education and Welfare, Public Health Service, Environmental Health Service, National Air Pollution Control Administration, Raleigh, N.C., Publication No. AP-73, August, 1970, p. 3 (percentages added).

TABLE II

Nationwide emissions for the modes of transportation, by types of contaminants, 1968

Modes	Carbon monoxide (Amt)	(%)	Particu- lates (Amt)	(%)	Sulfur oxides (Amt)	(%)	Hydro- carbons (Amt)	(%)	Nitrogen oxides (Amt)	(%)
Motor vehicles										
Gasoline	59·0	92·5	0·5	41·7	0·2	25·0	15·2	91·6	6·6	81·5
Diesel	0·2	0·3	0·3	25·0	0·1	12·5	0·4	2·4	0·6	7·4
Aircraft	2·4	3·7	N(a)	—	N(a)	—	0·3	1·8	N(a)	—
Railroads	0·1	0·2	0·2	16·7	0·1	12·5	0·3	1·8	0·4	4·9
Vessels	0·3	0·5	0·1	8·3	0·3	37·5	0·1	0·6	0·2	2·5
Non-highway use of motor fuels	1·8	2·8	0·1	8·3	0·1	12·5	0·3	1·8	0·3	3·7
Total	63·8	100·0	1·2	100·0	0·8	100·0	16·6	100·0	8·1	100·0

(in millions of tons)

Source: Data compiled from those contained in the *Nationwide Inventory of Air Pollutant Emissions, 1968, op. cit.*
(a)—Negligible.

carbons. This is shown in Table I having reference to the nationwide emissions of selected contaminants by sources for 1968.

Transportation accounted for 63·7% of the total emissions of carbon monoxide; 51·9% of the emissions of hydrocarbons; 39·3% of the emissions of nitrogen oxides; 2·4% of the emissions of sulfur oxides and 4·3% of the emissions of particulates. The total tonnage for all contaminants emitted by all sources amounted to 214·2 million tons of which 90·5 million tons or 42·3% were derived from transportation on a nationwide basis.

Motor vehicles are the primary source of air

pollutants derived from transportation with the exception of sulfur oxides. This is brought out in Table II showing the nationwide emissions for the several modes of transportation, by types of contaminants for 1968.

The motor vehicles accounted for 59·2% of the total emissions of carbon monoxide emitted by all modes of transportation; 94·0% of the emissions of hydrocarbons; 88·9% of the emissions of nitrogen oxides; 37·5% of the emissions of sulfur oxides, and 66·7% of the emissions of particulates. The tonnage for all types of pollutants derived from motor vehicles was 83·1 million tons, which was equal to 91·8% of all pollutants emitted by all modes of transportation.

Of the emissions by motor vehicles, 81·5 million tons or 98% were emitted by gasoline-fueled motor vehicles. They were the primary source of air pollutants derived from motor vehicles by a relatively wide margin with the exceptions of sulfur oxides and particulates. The gasoline-fueled vehicles accounted for 92·5% of the emissions of carbon monoxide derived from motor vehicles; 91·6% of the emissions of hydrocarbons; 81·5% of the emissions of nitrogen oxides; 41·7% of the emissions of particulates and 25·0% of the emissions of sulfur oxides.

Atmospheric lead concentrations are increasing rapidly throughout the northern hemisphere including the United States.[2] Dr. Chow stated in a paper presented before the American Chemical Society, that an analysis of the snow in Greenland revealed lead concentrations 500 times greater than the natural level. Lead particulates in the atmosphere over San Diego, California are said to be increasing at a rate of 5% a year and on winter days may reach as much as 80% of the safety level set by government health agencies. Some 96% of the atmospheric lead particulates are derived from gasoline-fueled engines in the United States, therefore the removal of lead from gasoline would go far in ridding the air of this contaminant.[3]

FEDERAL AIR POLLUTION LEGISLATION RELATING TO AUTOMOTIVE EMISSIONS

A review of the air pollution control legislation, enacted by the Federal Government, reveals the growing public concern with respect to automotive emissions as sources of air contaminants. It also indicates the need for the formulation of criteria and emission standards as a basis for effective control in protecting the public health and welfare.

The *Clean Air Act of 1963*[4] Section 6, recognized that emissions from motor vehicles were of growing importance as a source of air contamination. It instructed the Secretary of Health, Education and Welfare to encourage the automotive and fuel industries to continue their efforts in developing devices and fuels to prevent the emission of harmful pollutants from the exhausts of motor vehicles and to maintain liaison with them. Provision was made for the appointment of a technical committee which was to meet periodically for the purpose of evaluating the progress in this connection and recommending relevant research programs.

The Secretary was to file with Congress a report, within 1 year following the passage of the act and semi-annually thereafter, as to measures taken for controlling automotive exhausts and for improving fuels. Furthermore, the report was to include the occurrence of pollution caused by automotive emissions, research progress relating to devices and fuels, criteria as to the degree of exhaust pollutants, the effects of improved fuels and recommendations for additional legislation. No provision was made for the establishment of standards applicable to automotive emissions. The primary purpose of the act was to provide for cooperation between the Federal and State governments as a basis for preventing and/or controlling automotive exhaust emissions which were injurious to the public health and welfare.

The *Motor Vehicle Air Pollution Control Act of 1965*[5] came more to grips with the problem of automotive exhaust emissions. Its primary objective was providing for the establishment of emission standards applicable to new motor vehicles and new motor vehicle engines. The Secretary was made responsible for the prescription of practical emission standards after giving consideration to their technological feasibility and costs.

The Clean Air Act of 1963 which under subsection 3(a) instructed the Secretary to establish national research programs for the prevention and control of air pollution did not include those pertaining to automotive emissions. This was corrected by the 1965 act which amended this subsection by adding paragraph (5) providing for accelerated research and development programs to determine ways and means of controlling hydrocarbon emissions caused by evaporation of gasoline in carburetors and fuel tanks; emissions of oxides of nitrogen and aldehydes from gasoline-powered and diesel-powered vehicles. The programs also considered the development of improved low-cost technique for reducing emissions of oxides of sulfur caused by combustion of sulfur-containing fuels.

Manufacturers of these vehicles and engines were required to maintain records and to allow access to them by authorized persons and to permit such records to be copied by them. Failure on the part of manufacturers to report or provide information as requested by the Secretary was also contrary to the law. In addition, it was made unlawful to render inoperative any devices or elements installed on or in new motor vehicles and new motor vehicle engines prior to their sale or delivery to the ultimate purchasers.

New motor vehicles and new motor vehicle engines could be exempted from the requirements of the act for research purposes, investigations, studies, demonstrations and training, for reasons of national security and protection of the public health and welfare. In addition, the Secretary was empowered to refuse admission into the United States of new motor vehicles and new motor vehicle engines which did not conform to the standards set by the act. New motor vehicles and new motor vehicle engines, designed for export and so labeled and tagged, were exempt from regulation.

The Secretary was given authority to issue certificates of conformity to manufacturers of new motor vehicles and new motor vehicle engines which met the requirements set by him in accordance with the act. The duration of these certificates were for periods prescribed by him, but in no case for less than one year. Manufacturers of such motor vehicles and engines were required to keep records, report and provide information.

Congress by the *National Emission Standards Act of 1967*[6] preempted control of emissions from new motor vehicles and new motor vehicle engines, which were detrimental to or likely to be detrimental to the health and welfare of the public. The emission standards to be prescribed were to be applicable only to automobiles manufactured during and after the model year 1968. The failure to provide for emission standards applicable to earlier models and engines has been and continues to be a retarding factor in achieving more rapid and effective control over automotive emissions.

No state or political subdivision thereof could adopt or attempt to enforce emission standards for new motor vehicles and new motor vehicle engines which were subject to the federal act. Neither could they require certificates, inspection or other approval of such vehicles and engines as conditions for their initial sale, titling or registration.

The act authorized the Secretary, after notice and hearing, to waive the requirements of the act with respect to any State which had adopted standards, other than crankcase standards, for the control of emissions from new motor vehicles and new motor vehicle engines prior to March 30, 1966. No waiver could be granted, if he found that a State did not require standards to meet compelling and extraordinary conditions or that its standards and enforcement procedures were inconsistent with those of the federal government.

The Secretary was further authorized to designate fuels, which after the date or dates prescribed by him, could not be introduced into interstate commerce except under stated conditions. The manufacturers or processors of such fuels were required to provide him with information in keeping with the requirements, and any additives contained in them were to be registered with him.

The *Clean Air Amendments of 1970*[7] include important changes relating to the control and prevention of air pollution, caused by automotive vehicle emissions. The Administrator[8] is required to prescribe and revise from time to time emission standards applicable to new motor vehicles and new motor vehicle engines which endanger or may endanger the public health and welfare. These standards are to remain in force throughout the useful life of the vehicles and engines to which they apply. The useful life of light duty vehicles and light duty engines is stated as 5 years or 50,000 miles, whichever occurs first. For other motor vehicles and engines, it shall not be less than that for light duty motor vehicles and engines, unless the Administrator prescribes a longer period and a higher mileage requirement. Emission standards promulgated by the Administrator become effective after allowing time for the development and application of the necessary technology, giving appropriate consideration to the cost of compliance within such period.

The emission standards applicable to *carbon monoxide* and *hydrocarbons*, emitted from light duty vehicles and engines, manufactured during or after the model year 1975, shall contain standards which require a reduction of at least 90% from emission standards for these contaminants applicable to light duty vehicles and engines, manufactured in the model year 1970. There is a different provision for emissions of *oxides of nitrogen*. In this instance, it is required that the emission standards for this pollutant applicable to light duty vehicles and engines, manufactured during or after the model year 1976, shall contain standards which require a reduction of at least 90%, from the average of the emissions of this pollutant actually measured from light duty vehicles and engines manufactured during the model year 1971, which are not subject to Federal and State emission standards for oxides of nitrogen.

For the first time, a time-limit is set for the promulgation of emission standards and measurement techniques on which they are based. If they are not prescribed prior to the effective date of the act as amended, they shall be prescribed by regulation within 180 days thereafter.

A manufacturer of motor vehicles and engines may file with the Administrator a request for suspension of the standards for carbon monoxide and hydrocarbons applicable to motor vehicles and engines, manufactured during the model year 1975 at any time subsequent to January 1, 1972. If the petition is granted, the suspension cannot be for more than one year and interim standards shall be prescribed. The Administrator is required, after public hearing, to give his decision with respect to any suspension within 60 days. Requests for suspension of emission standards applicable to oxides of nitrogen, may be filed at any time after January 1, 1976. Here also, the suspension, if granted, cannot be for more than 1 year and the Administrator must render his decision within 60 days after the public hearing. Such suspensions are to be granted only when they are deemed essential to the public health and welfare.

The Administrator is empowered to enter into arrangements with the National Academy of Sciences to carry out a study as to the technological feasibility of meeting emission standards prescribed by him. He shall request from the Academy semi-annual reports as to the progress of the study to be submitted to him and to Congress, beginning not later than July 1, 1971 and to continue until the study is completed.

Certificates of conformity may be issued by the Administrator for emission control systems which have been incorporated in or on new motor vehicles or new motor vehicle engines, submitted to him for approval and after they have been tested adequately. These certificates can not be issued for periods in excess of one year. If a manufacturer is not satisfied with the decision of the Administrator, he may request a hearing to determine if the tests were carried out properly, but the suspension or revocation is not stayed by such action. If the manufacturer objects to the findings of the hearing, he may file a petition with the United States Court of Appeals for redress within 60 days of the issuance of the findings.

MOTOR VEHICLE EXHAUST EMISSION STANDARDS

Table III shows the variations in emission standards, both current and projected for light duty automotive vehicles and engines for the years 1968–1977, inclusive:

TABLE III

Current and projected emmission standards for automobiles and light duty trucks (6000 lb. GVW or less), 1968–1977

Model year	Test† procedures	Exhaust emissions, g/mile			Evaporation GM/test
		Hydrocarbons (HC)	Carbon monoxide (CO)	Nitrogen oxides (NO$_x$)	
1968–69‡	FTP	(275 ppm)	(1·5 vol. %)	NR	(—)
1970	FTP	2·2	23	NR	NR
1971	FTP	2·2	23	NR	6
1972	CVS	3·4§	39§	NR	2
1973	CVS	3·4	39	3·0	2
1974	CVS	3·4	39	3·0	2
1975	CVS	0·41††	3·4††	3·0	2
1976–77	CVS	0·41	3·4	0·4††	2

Source: From an unpublished study by A. Richard Schleicher, Senior Analyst Environmental Studies Center, Research Triangle Institute.

NR—No requirement.

†Federal Test Procedure (FTP) measures exhaust concentration, uses a 7 mode-7 cycle driving cycle.

(—)Constant Volume Sampler (CVS) measures true exhaust mass, uses a nonrepeating 1372 sec driving cycle (closed, self-weighting).

‡The 1968–69 standards are expressed as portions of exhaust gas, parts per million (ppm) or volume per cent.

§The larger numbers for HC and CO standards beginning 1972 are due to the fact that the CVS procedure gives larger readings than FTP. On an equal test procedure 1972 standards are more stringent than 1971 and do not represent a relaxation of previous requirements.

††1975–1977 HC and CO requirements and 1976–77 NO$_x$ requirements based on the Clean Air Amendments of 1970. Official definition of standards was published on 7-2-71.

The first federal emission standards were precribed in 1966 and were applicable to new motor vehicles and new motor vehicle engines, manufactured during the model year 1968. The standards were based on engine displacement and covered emissions of carbon monoxide in volume percentage and of hydrocarbons as so many parts per million. They represented reductions of about 50% for carbon monoxide and 70% for hydrocarbons when compared with emissions from uncontrolled automotive exhausts. The standard for emissions of hydrocarbons from crankcases was designed to achieve a 100% reduction, which required sealing of the crankcases and recirculating the unburned hydrocarbons for recombustion.

If the standards set in 1966 and applicable to 1968 model year automobiles are converted to grams per vehicle-mile, their equivalents are 33 g for carbon monoxide and 3·2 g for hydrocarbons. If emissions of these gases through the exhausts of uncontrolled vehicles prior to this time, are converted on the same basis, the equivalents are 71 g per vehicle-mile for carbon monoxide and 9·7 g per vehicle-mile for hydrocarbons.[9]

An estimate, based on the year 1968, sets the total quantity of pollutants, exclusive of carbon monoxide, emitted by automotive vehicles and engines at 80 million tons per year and assumes that this would be reduced to 76 million tons by 1969.[10] Furthermore, if more effective emission standards were prescribed and enforced, the tonnage might be lowered to 55 million tons by 1985. Thereafter, unless stricter control measures were adopted, the tonnage would rise because of the increase in vehicle ownership and vehicle-miles traveled.

The emission standards applicable to the model years 1970 and 1971 automobiles and light duty trucks are based upon the Federal Test Procedure and are expressed in grams per vehicle-mile. The standard for carbon monoxide is 23 g per vehicle-mile and for hydrocarbons 2·2 g per vehicle-mile. No requirement is given for nitrogen oxides for those years. The standards for the model years 1972 to 1974 inclusive, are based on the Constant Volume Sampler procedure utilizing cold-start mass testing, designed to simulate urban driving conditions. The larger numbers shown for these years therefore do not represent a relaxation of requirements for carbon monoxide and hydrocarbons. In fact, the standards are more stringent than those for previous years. The standards for 1976 and 1976–77 are based on the Constant Volume Sampler proce-

TABLE IV
Current and projected emission standards for heavy duty trucks (6000 lb GVN and over), 1967–1977

| Model year | Test procedure | Exhaust emissions‡ | | | | | | Evaporation GM/test | Smoke %‡‡ obscure |
| | | Conc., ppm or vol. % | | | Mass, g/bhph | | | |
		HC (ppm)	NOₓ	CO (vol. %)	HC + NOₓ	CO		
Gasoline engines†								
1967–69	NR	NR	NR	NR	—	—	NR	—
1970–71	Eng. Dyn.	275	NR	1·5	—	—	NR	—
1972	Eng. Dyn.	275	NR	1·5	—	—	NR	—
1973–74	Eng. Dyn.	275	NR	1·5	—	—	NR	—
1975–77	Eng. Dyn.	—	—	—	5	25	10	—
Diesel engines§								
1967–69	—				NR	NR		NR
1970–74	—				NR	NR		20–40
1975–77	Eng. Dyn.				5	25		20–40

Source: From an unpublished study by A. Richard Schleicher, Senior Analyst, Environmental Studies Center, Research Triangle Institute.
NR—No requirement.
†HEW engine dynamometer test cycle (steady 2000 rpm, various loads).
‡Concentrations are expressed on a volume portion basis through 1974 (parts per million (ppm) or volume per cent). After 1974, a mass basis of grams per brake horsepower (bhp) hour is used.
§EMA engine dynamometer test cycle (various stabilized speeds and loads).
††HEW engine dynamometer test—acceleration and lugging modes.

dure involving cold/hot weighting mass testing, which yield different values for the requirements relating to the emission of contaminants.[11]

Table IV shows the variations in emission standards, both current and projected for heavy duty trucks, for gasoline and diesel engines covering the years 1967–1977, inclusive. The standards for gasoline-engines are based upon engine dynamometer test procedure and are the same for the model years 1970–1974. The requirements are expressed on the basis of percentage of volume for carbon monoxide (1·5) and parts per million for hydrocarbons (275). No standards are given for nitrogen oxides. After 1974, the standards for these gases are based on mass stated in terms of grams per brake horsepower hour. It is to be noted that starting with the model year 1970, there are opacity requirements for smoke, ranging from 20% to 40% obscurity. Starting with the model year 1975, standards, based on engine dynamometer test procedure and stated in grams per brake horsepower hour, will be applicable to gasoline and diesel heavy trucks for hydrocarbons and nitrogen oxides (5) and carbon monoxide (25). By that year, gasoline engines on heavy duty trucks will be subjected also to evaporation requirements of 10 g per actual test.

PRESERVATION OF THE NATURAL BEAUTY OF THE LANDSCAPE CONTIGUOUS TO HIGHWAYS

The growing awareness of the need for maintaining and preserving the natural beauty of the Nation's landscape contiguous to highways and for more effective utilization of land bordering such avenues of automotive traffic brought about the enactment by Congress of the *Highway Beautification Act of 1965*.[12] Title I declares that the creation and maintenance of outdoor advertising signs, displays and devices in areas adjacent to the interstate and primary systems of highways are to be controlled in order to promote safety and recreational values for public travel and to preserve the natural beauty of the landscape.[13] It provides that on and after January 1, 1968, the Federal-aid highway funds, which are apportioned to any State not having made provisions for the effective control of the erection and maintenance of such means of advertising, located within 600 ft of the nearest edge of the right-of-way and visible from the main travel way, shall be reduced by 10%. Compensation is

also provided for the removal of such outdoor advertisements which were lawfully in existence on the effective date of the act; lawfully in existence along highways which became part of the interstate and primary systems on or after that date or before January 1, 1968, and lawfully erected on or after that date. The Federal government's share of such compensation is limited to 75%.

Provisions are set forth in Title II of the act for the control of outdoor junkyards, inclusive of "automobile graveyards" located along the interstate and primary highway systems, within 1000 ft of the nearest edge of the right-of-way and visible from the main travel way.[14] Federal funds, appropriated to any State, may be reduced by 10%, if it has not established effective control measures applicable to such undesirable sites. The Federal government's share of the cost of screening and landscaping in relation to these unsightly business sites is limited to 75%. Compensation is provided for owners where such operations are relocated or disposed of. In such instances, the cost of the relocation or disposal is covered up to 75% by the Federal government.

Title III of the Highway Beautification Act, relates to the landscaping and the enhancement of scenic areas adjacent to highways.[15] The Secretary of Transportation may approve, as part of the cost of construction of Federal-aid highways, the cost of landscaping and roadside development of publicly owned and controlled rest and recreation areas and sanitary and other facilities reasonably necessary to accommodate the traveling public. An amount equal to 3% of the funds appropriated to a State for Federal-aid highways for any fiscal year shall be allocated to that State for such purposes without being matched by the State. No part of the Highway Trust Fund shall be diverted for these purposes.

The Department of Transportation responded in 1969 to the nationwide concern about the impact of highways upon the environment by establishing an Office of the Assistant Secretary for Environment and Urban System.[16] This action provided for a bridge between transportation objectives and fundamental environmental goals. The Federal Highway Administration in the Department of Transportation has in recent years sought ways and means whereby the Federal-aid highways can be made more compatible with the environments through which they pass. It is not merely concerned with minimizing the possible adverse effects but the utilization of highways as a positive force in attaining social and economic environmental objectives. About 12% of the total costs of the Federal-aid highways projects is devoted to environmental aspects, which for the fiscal year 1969 amounted to approximately $580 million.[17]

Organizationally, every office of the Bureau of Public Roads is concerned with the environment relative to Federal-aid highways. During the fiscal year 1968, there was created within the Bureau, the Environmental Development Division of the Office of the Right-of-Way and Location, to help insure that the environment of affected areas will be studied and fully considered in the future location, design and construction of all Federal-aid highways. This division is responsible for the protection and enhancement of human values and resources of the landscape along the rights-of-way of these highways. It is staffed by an interdisciplinary team made up of urban planners, architects, sociologists, economists, real estate appraisers, highway engineers and others with related expertise.

MOTOR VEHICLES AND "NUISANCE" NOISE

Noise pollution is receiving considerable attention and increased emphasis is being given to "nuisance" noise, caused by automotive vehicles, especially those operating in or near urban centers. One of the major sources of noise is the heavy duty trucks moving along the Nation's highways and streets.

In general, traffic noise radiating from freeways, mid-town shopping centers and apartment sections of cities publicly disturb more people than any other source of outdoor noise.[18] It is estimated that a single truck moving at expressway speeds, often generates sound levels exceeding 90 dB (decibels) and a long line of truck traffic may generate levels as much or more than 100 dB. These sound levels may prove detrimental to the physiological health of persons who are continually subjected to them. It has been stated that half of the people who are exposed to extremely heavy city street traffic may suffer some loss of hearing. The skin may become flushed, stomach muscles may be constricted and tempers may be shortened, when people are exposed to as much as 90 dB.

In response to the need for noise abatement, there is incorporated in the Clean Air Amendments of 1970, Title IV which is cited as the *Noise Pollution and Abatement Act of 1970*.[19] This title provides for the establishment by the Administrator of an

Office of Noise Abatement and Control within the Environmental Protective Agency. Its major responsibilities are to conduct a full and complete investigation of noise and its effects upon the public health and welfare and to identify and classify its causes and sources. This investigation is now in progress.

RESEARCH AND EXPERIMENTATION

The automotive industry is lending its cooperation through research and experimentation, relating to motor vehicle engine modifications and the development of devices and systems to reduce emissions of injurious contaminants into the atmosphere. This is being done even though the industry has certain reservations as to the practicality of the emission standards and its ability to conform to them by 1975 and 1976. It is doubtful that new motor vehicle propulsion systems will be developed, capable of being produced on a mass production basis, as replacements for the gasoline-fueled internal combustion engines prior to 1975 or 1976. There are several types of power systems under investigation at the present time, including the Rankine cycle engine, gas turbine, heat engine impact/electric hybrid, heat engine/fly wheel hybrid, all electric engines, stratified charge engines and advanced design diesel engines.[20]

The oil companies are engaged in extensive research and experimentation in the development of more adequate fuels and additives which will reduce the quantities of air contaminants emitted from the exhausts of automotive vehicles.[21] It is estimated that they have spent as much as $250 million on air pollution research during the last 10 years and that, the American Petroleum Institute is currently spending around a million dollars annually on problems associated with automotive air pollution. Illustrative of the costs involved in such undertakings, is the removal of lead additives from gasoline. This will probably add about $1·5 billion annually to the motorists' fuel costs and $425 billion in additional construction of refineries. Environmental reconstruction is largely a problem of economics, involving trade-offs based upon opportunity costs.

It is believed that much can be accomplished in reducing the impact of automotive transportation upon the environment through the more effective planning of cities.[22] There appears to be considerable correlation between spatial planning, meteorological elements, topographical features and the impact upon the environment by automotive transportation. The structural profiles of cities in conjunction with other elements affect the degree of concentration of contaminants in different places at different times.[23] In addition, the prevailing conditions under which motor vehicles are operated have a determining influence upon their impact on the environment.[24]

THE POTENTIAL THREAT OF AUTOMOTIVE TRANSPORTATION TO THE ENVIRONMENT

The need for the formulation and the rigid enforcement of effective emission standards become more apparent when viewed in the light of automotive registrations, the annual output of motor vehicles in the United States, and their respective growth trends.[25] These show why motor transportation is one of the major sources of environmental contaminants and the threat which it holds for the environment, if adequate measures for its control are not instituted and applied. There were in 1970 about 105 million automotive vehicles registered in this Nation, consisting of 87 million automobiles and 18 million trucks and buses. It possesses approximately 49% of the world's automobiles and 37% of its trucks and buses. Each year there are produced more than 10 million new motor vehicles and some 7 million old vehicles are discarded, leaving a net addition of 3 million motor vehicles.

In 1969, the number of motor vehicle-miles exceeded 1015 millions which were divided about equally between rural and urban areas. In accumulating this enormous mileage, over a road and street system of 3·5 million miles, some 87 billion gallons of gasoline were consumed—partially consumed—resulting in huge quantities of pollutants being emitted through evaporation from carburetors and gasoline tanks and emissions from exhaust pipes. The largest part of these pollutants were emitted through millions of tailpipes, located close to the surface, making them dangerous to the health of persons, especially those residing in congested areas of cities.

In addition, motor travel in American cities has been increasing at a faster rate than the urban population. The population growth for the period 1950–1960 was 18% as compared with 52% for urban motor travel for the same period.[26] It is anticipated that urban areas will absorb all of the 70 millions increase in population from 1960 to 1980 and that 50 millions of this increase will live in surburban areas which should make for greater dependence upon automotive transit in these areas. Stated in a different way, it is estimated that the population of the United States will increase between 1960 and 1980 by 39% as compared with increases of 64% in automobile registrations, 61% in truck registrations and 85% in motor vehicle travel-miles, three quarters of which will likely take place in urban areas.[27]

It is imperative that effective environmental policies be formulated and implemented at once and that long range environmental goals replace shortrun expediencies. The disastrous impact of automotive transportation on the environment must be avoided, if the health and welfare of the American people are to be adequately safeguarded.

REFERENCES

1. *Public Law* 91–190 (January 1, 1970).
2. Tsaihwa J. Chow (cited in Editorial), *Transportation & Distribution Management*, p. 2. (May, 1970).
3. *Development of Systems to Attain Established Motor Vehicle and Engine Emission Standards*, Report of the Administrator of the Environmental Protection Agency to the Congress of the United States, pp. 4–8. (September, 1971).
4. *Public Law* 88–206 (December 17, 1963) Section 6, "Automotive Vehicle and Fuel Pollution".
5. *Public Law* 89–272 (October 20, 1965) Title II, section 201 provides that this title may be cited as the "Motor Vehicle Air Pollution Control Act".
6. Title II, Air Quality Act of 1967, *Public Law* 90–148 (November 21, 1967), Section 201 provides that this title may be cited as "National Emission Standards Act".
7. *Public Law* 91–604 (December 31, 1970) "Motor Vehicle Emission Standards and Regulation of Fuels".
8. The Administrator of the Environmental Protection Agency which replaced the National Air Pollution Control Administration (December 2, 1970).
9. John T. Middletown and Wayne Ott, "Air pollution and transportation" *Traffic Quarterly* 22 (2), 179 (April, 1968).
10. Tom Alexander, "Some burning questions about combustion" *Fortune* 168 (February, 1970).
11. "Development of systems to attain established motor vehicle and engine emission standards" *op. cit.* 2–4.
12. *Public Law* 89–285 (October 22, 1965).
13. Title I, Section 101, Section 131 of Title 23. United States Code is revised to read as follows: *Control of Outdoor Advertising*.
14. Title II, Section 201, Chap. I of Title 23, United States Code, is amended to add at the end thereof the following new sec. 136 *Control of Junkyards*.
15. Title III, Section 301 (a). Section 319 of Title 23, United States Code is revised to read as follows: Sec 319. *Landscaping and Scenic Enhancement*.

16. *Department of Transportation Fourth Annual Report* p. 33 (1970).

17. *Highway Environmental Reference Book*, p. 3. U.S. Department of Transportation/Federal Highway Administration, Washington (November, 1970).

18. Robert H. Haskell, "Transportation and the environment" *Transportation and Distribution Management*, 30 (April, 1970).

19. *Public Law* 91–604. *op. cit.*, Title IV, *Noise Pollution*, Section 401 is cited as the "Noise Pollution and Abatement Act of 1970".

20. Development of Systems to Attain Established Motor Vehicle and Engine Emission Standards, *op. cit.*, pp. 3–9.

21. Cullison Cady, "For cleaner air" *Highway User*, Part I, 14–17 (November, 1967); Part II 14–17 (December, 1967); and Part III, 16–19 (February, 1968).

22. C. Peter Rydell and Gretchen Scharz, Review Article: "Air pollution and urban form" (Hunter College, New York, Urban Research Center, 1967). See also: A. F. Bush, "Urban atmospheric pollution" *Civil Eng.* 35 (5), 66–68 (May, 1965).

23. "Reduction of air pollution potential through environmental planning" (Air Pollution Control Association, New York, N.Y. 1970). (Presented at the 63rd Air Pollution Control Association Meeting, St. Louis, Mo., June 14–19, 1970).

24. D. M. Teague, "Los Angeles traffic pattern survey" in *Vehicle Emission*, SAE Tech. Progress Series, Vol. 6 (Society of Automotive Engineers, New York, 1964). (Presented at the National West Coast Meeting, Society of Automotive Engineers, August, 1967).

25. Unless otherwise indicated the data with respect to the Automotive Industry have been obtained from *Automobile Facts and Figures, 1970.* (Automobile Manufacturers Association, Detroit).

26. *Urban Transportation Issues and Trends* (Automobile Manufacturers Association, Detroit, June, 1963) p. 1, Figure 1; p. 12, Figure 6 (percentages calculated from charts).

27. *Ibid.*, p. 21, Figure 14.

THE NEEDS OF COMMUNITIES AND
LEGISLATION OF THE ROAD TRANSPORT VEHICLE

ANTHONY DAVID FRANK BAMPTON

Perkins Engines Company, Peterborough, U.K.

Transportation systems all over the world have been traditionally geared to high operating efficiency and economy. This is what business operations and the customer in the street has demanded. We now see another need, rising from a social recognition of our past neglect of the environment in which we live and create for ourselves. A need which can only acquire acknowledgement through standards set by legislation to create a situation in which industry can compete equally. There is no doubt that without this spur, many technologies which have developed so rapidly in recent years would have remained dormant. However, in spite of these technological developments in materials and processes, the cost of transportation will certainly increase. The amount of increase depends upon the rationality of the legislators.

There is a very real need to establish basic criteria concerning the control of man's pollution of the environment from all sources. So far as the transportation industries of trading nations are concerned, criteria on a global basis are necessary in order that countries may have common ecologically and economically balanced standards. This may become possible through the recent proposals announced for a United Nations "Earthwatch Programme".

INTRODUCTION

The task of governments to legislate environmental standards for the community is not simple. No administration can simply aim to eliminate pollution, without first exploring the cause of its presence. In the case of pollution from industry, whether it is produced at the processing stage, generated by the end product itself during its life-time or by disposal of the product after its useful life is expended, the reason can invariably be traced to cost.

How much is a community prepared to pay for quality of life? To determine the cost of improvement to standards is generally not difficult. Industry can readily assess the cost incurred to reduce waste product. The difficulty comes in determining the improvement to be gained from any such reductions, and then representing this in measurable terms (cash).

A curve which is characteristic of this equation is shown in Figure 1, and may be taken to represent the problem of the motor vehicle. The benefit to the community of the motorized vehicle for passenger conveyance, provision of an economic goods transport system, its role in civil construction and agriculture are all well known. But what is the contribution of these appliances to the pollution problem, of which there is acute consciousness today—the danger to health of accumulated toxic emissions in city streets, the nuisance of diesel exhaust smoke, the damage to hearing and to the nervous system? The medical evidence available is not decisive. It is even less certain that a measure could be meaningfully described when pollution from natural sources is taken into account.

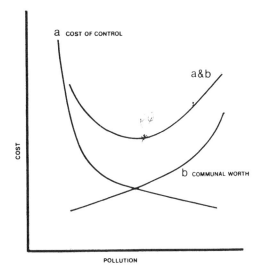

FIGURE 1 The cost of pollution and pollution control.

This then is the problem which faces the legislator, and which must be faced up to if the community is to have good and economically justified legislation.

WORLD LEGISLATION

The parts of the world in which there is a conscious-ness of the need for control of pollution and other damage caused by the motor vehicle, and currently introducing legislation on an intensive scale, are North America, Western Europe and the U.K. Other countries in which legislation is being, or shortly expected to be formed, are parts of Latin America, South Africa, Australia and Japan.

Until recently there has been no co-ordination of the legislation and the standards introduced domestically by separate countries and states, with the result that trading is impaired. Transcontinental transport operators face border restrictions if vehicles fail to comply with national regulations. Manufacturers for export markets are faced with a range of different standards that their products must meet, which impairs their efficiency of production and trading competitiveness. This situation stems generally from an uninformed politically parochial attitude. However, there is now an increasing degree of co-operation between some Governments to devise rationalized standards in legislation. This is a first step towards arriving at a balance between the economic and ecological benefits which can be made available to the community. The Economic Commission for Europe (E.C.E.) sets standards for voluntary adoption by the member countries. These standards generally form the basis for regulations for the European Economic Community (E.E.C.) nations. The Federal Government of the United States is setting, through its rule making body, the Environmental Protection Agency (E.P.A.), standards and regulations which will pre-empt much of the fragmented State law which exists now. Unfortunately, it appears that some of its legislation is being influenced too greatly by the pioneering work of the California State Legislature. There is a geophysical and climatic condition in this State which is almost unique and this should not dictate standards which are generally set.

Although there is harmonization activity going on within trading blocks and other neighbouring communities having common interests, there is no alignment now of environmental legislation on a global basis which embraces the contributions to pollution made by the motor vehicle.

CONTROL OF VEHICLE NOISE

The noise which reaches the bystander is a combin-ation of noises emanating from a variety of sources on the vehicle. These sounds occur at a wide range of frequencies from the different vehicle components. The widely accepted measure of these sounds is the decibel (dB). Also, because the ear varies in receptivity to sounds occurring at different frequencies, a weighting is applied to the measured values to compensate for this. Noise is therefore expressed, and the legislated standards are quoted

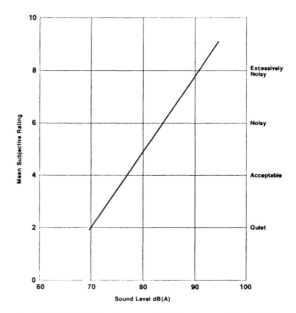

FIGURE 2   Relation between diesel truck sound level dB(A) and subjective rating.

in decibel "A" weighting units—dB(A). A relation-ship between subjective rating and dB(A) for trucks is shown[1] in Figure 2. For the purposes of comparing different levels of noise it is necessary to emphasize the exponential relationship that exists. A shift of 1 bel, or 1 decade on the decibel scale, in sound pressure level requires a ten-fold change in acoustic energy. This amounts to a two-fold change in the noise experienced by the observer. Alternatively, a reduction in noise of 3 dB requires a two-fold reduction in acoustic energy.

Possibly the most significant advance in legis-lation against the vehicle noise nuisance to third parties is that announced by the U.K. Government for enforcement in 1974—Table I.

These new noise limits represent a lower level of permitted noise in all vehicle categories than any other major piece of noise legislation with the possible exception of that in Switzerland. The

TABLE I
U.K. noise limits for road vehicles

Applicability	Current law (dBA)	Future law (dBA) 1974
Light goods vehicle or light passenger vehicle of not more than 3·5 tons G.V.W.	85	82
Any other goods vehicle of more than 3·5 tons G.V.W. and powered by an engine which does not exceed 200 b.h.p. (installed)	89	86
Any other goods vehicle of more than 3·5 tons G.V.W. and powered by an engine which does exceed 200 b.h.p. (installed)	92	89

Test procedure to B.S. 3425.

reduction of 3 dB(A) called for, represents an almost 25% reduction in subjective by-stander noise and requires the vehicle manufacturer to halve the existing level of acoustic energy emitted.

The U.K. Government has set for the longer term a target[2] of 80 dB(A) for the heavy lorry. This sets the manufacturer a very demanding objective which will require quite radical changes to current vehicle designs. Removal of the cooling system (fan and radiator) to some other less obtrusive part of the vehicle; improving muffling of the exhaust and air intake; and reduction of engine emitted noise by major structural redesign or enclosure systems are all areas for early attention. Reduction of tyre and transmission noise will probably also be necessary. The Department of the Environment has established a research programme to study these factors.

A study of current reciprocating diesel engine noise has been made[3] to find out if there is a particular type of engine which offers an advantage. An empirically derived relationship of noise with basic engine parameters was thereby established:

$$I \sim N^n \times B^5 \qquad (1)$$

I = noise intensity dB(A)
N = engine rotational speed
B = Cylinder bore
n = empirical value dependant upon engine type.

Shown in Table II are results calculated for a 225 hp engine.

Basic engine durability standards are observed in Table II by placing a limitation on engine piston speed and brake mean effective pressure. Stroke/

bore ratio is selected according to general practice for each engine configuration.

The importance of designing for minimum cylinder bore diameter can be appreciated from Eq. (1). This reflects the need to reduce the impulsive forces acting on the engine structure at a given mean effective pressure. However, in practice it is shown by the examples given in Table II for a particular horsepower requirement that the differences in noise attentuation of current engine forms is not great. There is little more than 4 dB(A) bare engine noise difference between the large bore low revving 6-cylinder four-stroke engine and the smaller bore but higher revving two-stroke. From the point of view of noise consideration alone, it is unlikely then that any one configuration of engine will have a particularly significant advantage over another.

For the more immediate requirement to comply with the regulation to be made law and applicable

TABLE II
Predicted diesel engine noise

Engine	Cylinder bore (in.)	Rated speed (rpm)	Bare engine noise (dB(A) at 3 ft)
8 cylinder	4·75	2400	103·7
8 cylinder turbocharged	4·25	2700	102·2
6 cylinder	5·375	2000	104·0
6 cylinder two-stroke	3·875	2750	101·1
6 cylinder two-stroke, turbocharged	3·50	3000	99·8

to vehicles for sale after 30 September 1974, manufacturers need to take action now if they are to introduce proven changes into their product lines. This does not apply to all manufacturers as some vehicles already meet the new standard. This legislation is aimed to bring the more offensive vehicles into line with the current best. The changes contemplated are interim to the longer term objective of the 80-dB vehicle, and will mostly be secondary modifications to existing vehicle forms. The sort of changes which may be expected are: smoothing of the combustion process to reduce impulsive loading on engine structure; detail attention to the design and choice of materials for external engine components; shielding of engines or engine compartments with acoustic barrier and absorbent material; and attention to engine mountings, cooling fan design, exhaust and intake systems. There is no standard treatment which can be applied to all vehicles and manufacturers must discover which is the offending area(s) requiring treatment on each vehicle model. The choice of solution and reduction in noise obtained then needs to be weighed against cost and the flexibility of application in manufacturing to meet customer and legislative demands in various marketing countries.

This legislation, coupled with regulations for smoke, has been estimated will cost £10 million per year, falling mainly on heavy lorries and buses which will often have up to £100 added to the cost and in some cases[4] over £200. A large part of this is attributable to the noise packages. This will add to the cost of transportation and must reflect the nuisance value of subjective noise experienced by the community.

The drivers of these vehicles must endure a higher level of noise for a prolonged period than the by-stander. In the majority of lorries today drivers are subjected to 85–90 dB(A) with some vehicles being even higher. Enforced restriction of this noise to a maximum of about 85 dB(A) can perhaps be expected in the U.K. within the next 3 years.

To have dwelt on the legislation for the U.K. and its effect, does not imply that other countries are inactive. Regulations are in force in many countries, but the standards they adopt do not reach those planned for this country.

EXHAUST EMISSIONS

The legislation for exhaust emission is concerned separately with particulates, and with certain toxic or oxidising gaseous elements. The emission of particulates is only really of concern in the case of the diesel engine, which issues free carbon as soot in the exhaust creating a nuisance. The gaseous elements which are found to be restricted in most current legislation are carbon monoxide (CO); hydrocarbons (HC); and oxides of nitrogen (NO_x), usually reduced to NO_2.

Much has been written in the popular press of the contribution of certain industries, power generation plant, motor vehicle etc., to the problem of atmospheric pollution. Unfortunately, many reports are made without due regard to the real problem of toxicity, and one more frequently sees reference to enormous tonnages of sulphur, carbon monoxide, particulates etc. California Air Quality Criteria establishes how much of each pollutant can safely

be tolerated in the atmosphere, from which weighting factors are derived:[5]

carbon monoxide	1
oxidants (hydrocarbon products)	27·9
oxides of nitrogen	52·2
oxides of sulphur	235
particulates	249

The effect which the application of these weighting values has on the role of some major pollution sources is shown in Table III. These are based upon 1965 tonnage figures from the U.S. Department of Health, Education and Welfare. Similar conclusions are drawn by other researchers.[6]

TABLE III
Rank order of pollution sources

Pollution source	Rank order percentage	
	Tonnage	Weighted
Industry	16	37
Power plants	14	36
Vehicle	61	12
Space heating	6	10
Refuse disposal	3	5

The major role of the motor vehicle as a contributor to atmospheric pollution on a weight basis is due to the relatively high emission of carbon monoxide from the gasoline engine. For each ton of fuel burned, about $\frac{1}{3}$ ton of CO is released into the atmosphere from an engine not subject to emission control. Present emission control in America reduces this by about two-thirds. In terms of toxicity per unit weight, CO is viewed as being quite a mild pollutant, as is the free carbon found in diesel engine exhaust. The motor vehicle contributes very little in oxides of sulphur and particulates which are considered to be the two major pollutants.

These weightings which are devised for California, may not be applicable to other parts of the world where the geophysical and climatic effects are not the same. The weighting given to the oxidants in hydrocarbon products and oxides of nitrogen for example, may be quite different. A photo-chemical reaction of elements in the exhaust, believed to occur principally between certain hydrocarbons and oxides of nitrogen in the presence of sunlight produces a smog which is retained in a static air stream over the California basin. Tokyo is a heavily trafficed city which has a similar climatic condition.

This glance in perspective does not suggest the relief of pressures to reduce pollution from any of the major sources, but it does assist in setting the priorities.

Control of Smoke

Regulations for the control of smoke from diesel engine road vehicles (weighing more than 6000 lb G.V.W.) were in force by the U.S. Federal Government for 1970 model year vehicles (see Table IV). This regulation requires a manufacturer to satisfy the standards set and obtain certification from the Environmental Protection Agency. This is accomplished through the satisfactory completion of a test programme with each engine type for which certification is required.

A similar form of regulation will be introduced into the U.K. on 1 October this year. Certificates of Conformity to the Standard are granted by British Standards Institution, which administers this process on behalf of the Department of Environment. The Standard known as B.S. AU 141a:1971 is very much in line with that of the Draft Proposal set by the Council for the European Economic Community. There should be no conflict therefore when Britain joins the Community. This is not so in the case of the U.S. Federal Standard which is quite different in both engine test procedure and emission level permitted from those of U.K. and E.E.C. The Federal Standard is generally found by manufacturers to be less demanding.

The permissible smoke limit set in the British Standard was devised from the judgement made by panels of independent and unbiased observers of drive-by vehicle exhaust conditions. The limit found objectionable by 50% of the observers was chosen as the target which industry should work towards. This limit is shown in Hartridge units of opacity in Figure 3. Unlike the rudimentary Standards which exist in some countries, it should be noted that provision is made in the Standard for the effect of engine capacity and operating speed (nominal gas flow). For comparative purposes the U.S. Federal limit is shown for the lugdown mode of operation (simulating driving on a gradient such that progressive engine speed reduction occurs at maximum load). This approximates to the engine steady state full load operation in the British test. It has also been necessary to correlate readings from the different opacimeters used, so that the Federal limits shown must be taken as approxi-

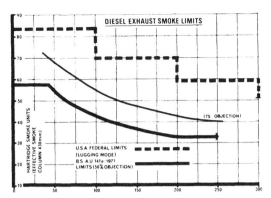

FIGURE 3   Nominal gas flow litres/sec ($\simeq$ rated hp).

mate. It is to be noted that allowance for engine capacity is made in the E.P.A. Standard in which the exhaust stack diameter (which defines the optical length of the smoke column) which should be used for a particular engine power range is specified. In spite of limited accuracy of the comparison the U.S. Federal level is one which many more than 50% of the British population would object to. The 75% objection level defined by the British Panel of observers is shown in Figure 3.

The standard which individual manufacturers in this country set for themselves before legislation was contemplated has varied. In some cases, therefore, the legislation has resulted in a defuelling and lowering of horsepower output. In others where a manufacturer has restricted horsepower to meet his already very high regard for smoke emission, he is able to increase horsepower and meet the national standard for smoke. This legislation therefore puts a brake on any tendency for worsening of the current situation, in addition to removing the worst offenders from the road rather than setting a lower national threshold limit. The action now taken at the manufacturing stage is sufficient for the general public to receive a noticeable reduction in nuisance and offence for a minimum cost. This will be further reinforced through observation by the vehicle operators of proper maintenance standards encouraged by more effective government imposed in-use regulations. These have been promised.

There is to be a tightening of the Federal Standard, announced by the Environmental Protection Agency, to be introduced on 1 January 1974. The expected limit of opacity is shown in Table IV with the existing limits. In this revision of legislation a limitation is also put on the puff of black smoke which is sometimes seen when a vehicle accelerates from rest and changes through the gears. This is particularly noticeable of the turbocharged engine in which the response of the turbocharger lags the throttle movement.

TABLE IV

U.S. Federal smoke limits

	Opacity limits—U.S. P.H.S. meter	
	---	---
Mode of operation	Current (%)	Expected 1 January 1974 (%)
Acceleration	40	20
Lug-down	20	15
Half-second peak	—	50

Even this revised Standard does not bring the lug-down levels down to the present U.K. limits shown in Figure 3, and would still be found objectionable by a large proportion of the observers used in establishing the British Standard.

Expected to occur on 1 October 1974 is the introduction by the Department of the Environment of a further requirement to the 1972 legislation. Engine manufacturers will be required to provide evidence that *all* engines produced for use on the road do meet the standard laid down. In effect, this asks for confirmation that the design standard which is proven in the Type Certification Test is being produced consistently in manufacturing production. Since in any mass producing industry a tolerance is allowed on the reproduceability of articles made, it is expected that the final product will perform to its objective within an allowed band of deviation. Manufacturers must therefore ensure that the performance claimed for the product is the minimum in order to satisfy the 1974 revision. This could bring about a review of manufacturing control procedures where a manufacturer has not been accustomed to expressing a minimum performance by his product, but has perhaps used mean achievement. The general public will again see this as a further step in the reduction of smoke on the roads.

Gas Emission

The United States is without doubt the most active nation in controlling gas emissions from road vehicles. First controls were introduced in 1968 for the "light duty" automobile. This embraces in the

definition the private motor car. The standards have since been tightened and there are plans for further tightening, which by 1985 it is forecast by the Automobile Manufacturers Association will hold the weight of pollution emissions down to 75% of the pre-1968 uncontrolled level, in spite of vehicle population growth.

It has been announced by the Environmental Protection Agency that emission controls will be extended in the U.S. to include the "heavy duty" motor vehicle on 1 January 1974. Although just at this point in time there is no clear statement of the emission levels which are to be met, it is expected that diesel engine manufacturers for the truck market will need to do some development work to satisfy the standard. The reason for this doubt concerning the standard, is brought about by an E.P.A. announcement made in February this year of deferment of the introduction of emission regulations from 1 January 1973 because "inadequate lead time for compliance with the 1973 Standard made the deferment necessary". The E.P.A. did, at the same time, abandon its own proposed emission limits. It is expected that the Environmental Protection Agency will adopt limits similar to those to be introduced by the State of California on 1 January 1973. These are shown in Table V.

TABLE V

U.S. heavy duty gas emission limits

Emission (g/hp/h)	California 1 January 1973	Federal 1 January 1973 (now withdrawn)
CO	40	7·5
HC	—	3·0
NO_2	—	12·5
HC + NO_2	16	—

Although many manufacturers found difficulty in complying with the Federal limit of 7·5 g/hp/h of CO when using the prescribed test cycle, the HC and NO_2 requirement is also not easily met by many current diesel engines. The withdrawal of this Standard and its replacement by a Standard similar to that to be introduced in California, does not then substantially ease the problem for certain types of diesel engine, even though a combined level of HC and NO_2 is specified. The California limit of 40 g/hp/h for CO is the limit established jointly for both gasoline and diesel powered vehicles, which the diesel can meet very easily.

This legislation then does partially but effectively achieve the objective of constraining the diesel engine manufacturer and forcing a contribution to the protection of the environment. If, however, the Environmental Protection Agency does adopt the 40 g/hp/h limit for CO, this will represent a major sacrifice of this principle, and would unwillingly concede to the diesel manufacturers viewpoint that the contribution by the diesel to atmospheric pollution by CO is small and that punitive legislative control is unnecessary.

It is proposed to revise and introduce in 1975 and 1976 new standards of emission control by California State and by Federal Authority. The Californian proposal at present contemplated is 25 g/hp/h for CO and 5 g/hp/h combined HC and NO_2. This level of combined emission will present large sections of the industry with major difficulty. If persisted in, this could put the American road vehicle industry seriously out of step with the rest of the world, as in fact their policy towards the private motor car is already doing.

Discussions for the legislation of gas emission is only at the preliminary stages in Europe. It is felt there will be no great urge to follow the lead set in America without first waiting to see the effect on the economy of their severe approach, and without

making a careful study of the environmental needs for control. It is unlikely there will be any enforcement of standards in Europe before 1975.

There is a very real need to establish basic criteria concerning the control of man's pollution of the environment from all sources. As regards the problems engendered by the development of the motor vehicle, criteria on a global basis are necessary in order that countries may have common ecologically and economically balanced standards. This may become possible through the recent proposals announced for a UN "Earthwatch Programme".

REFERENCES

1. G. H. G. Mills and D. W. Robinson, "The subjective rating of motor vehicle noise" *M.I.R.A. Report No. 1961/3* (April 1961).
2. *The Protection of the Environment—The Fight Against Pollution.* H.M.S.O. Cmnd. 4373 (May 1970).
3. T. Priede, "Noise in engineering and its effect on the community" *Automotive Engineering Congress, Detroit, Mich. S.A.E. Paper No.* 710061 (11–15 Jan. 1971).
4. "Noise, smoke and power/weight ratios—Transport Ministers decisions on vehicle regulations" *Dept. of the Environment Press Notice* 216 M (7 Oct. 1971).
5. Robert F. Sawyer and Lawrence S. Carelto, Editorial of *Automotive Industries* (1 Oct. 1971).
6. Lyndon R. Babcock, Jr. and Niren L. Nagada, *Chemical and Engineering News* (10 Jan. 1972).

LEGISLATION AND THE MOTOR CAR

H. J. C. WEIGHELL, Wh. Sch., C.Eng., M.I.M.E., M.I.P.E.

Director, Legislation and Technical Liaison, Chrysler International SA,
Technical Centre, Whitley, Coventry CV3 4BG, U.K.

Vehicle legislation for safety and emission control is led by the USA where half the world's cars and one-third of the world's road deaths are found. The 1975/6 standards will cause confrontations between industry and government in the U.S.A. and have costly implications for the American people. The rest of the world is affected by what happens in the U.S.A. but for language and other reasons fragmented laws are normal with conflicting requirements for manufacturers who supply the many markets. The greatest force for harmonization of legislation may be the enlarged Common Market including, as it does, the four major European Car manufacturing countries. The Common Market Commission may displace WP 29 of the United Nations as the leading law-making body. Type approval is the compliance system used in Europe, but the growth in scope of safety and other legislation is bringing into focus a weakness of the system which is reducing production efficiency to an alarming degree. American style self-certification is not acceptable to European governments.

1. INTRODUCTION

Ever since the British requirement that motor vehicles should be preceded by a man carrying a red flag, it has been apparent that the blessings of motorized transport should not be enjoyed without restriction. When horse power was replaced by Horsepower, man had at his command greater forces than ever before and, in the motor car, these could be unleashed by the slight pressure of his foot and, if directed inexpertly, could cause untold damage.

In the modern family car, pressure on the accelerator is multiplied by the engine about 100 times to accelerate the car. At 30 m.p.h. that car and its occupants weighing about 1·25 tons have a kinetic energy of 38 ft tons. Impact with another object must have dire consequences. The proof is in the accident figures. Over 150,000 people were killed last year on the roads of the world: 55,000 in the U.S.A., 7700 in the U.K., 16,200 in France and 17,900 in W. Germany.

Man's strongest instinct is self-preservation and, in the modern societies of developed countries, this often manifests itself as laws intended to reduce the toll of death and injury on the roads.

To begin this review of the relationship between legislation and the motor car, I propose to look at first the United States of America whose roads carry about 100 million cars or half the total world car population.

2. THE AMERICAN SCENE

Not surprisingly, the country with the highest annual number of road deaths has reacted most aggressively to control the slaughter. Despite its passionate concern for individual liberty and entrepreneurial freedom, the U.S.A. has been the most active in passing laws to coerce the manufacturer into providing automobiles which will protect the occupant against the mistakes of others as well as himself. Their declared object is to eliminate road deaths and reduce human suffering and the destruction of property. The American law-making action has identified motor cars, and to a lesser extent, other road vehicles as the agents of death and destruction, and Congress acted with breakneck speed when it established a permanent administrative body with a multimillion dollar budget to regulate the safety aspects of motor vehicles.

The National Highway Traffic Safety Administration (NHTSA) has the authority to make safety standards which must be met by vehicle manufacturers on new vehicles produced after established dates. Provision is made for information and argument from all concerned bodies to be laid before the Administrator after proposed regulatory measures have been published in the Federal Register. All submissions for mitigation of the proposed rules are public and a limited period of time is provided for submission and subsequent consideration, thus ensuring that rules cannot be delayed for more than a few months even though they may be amended. The duty of the Administration is to act in the public interest whatever effect this may have on industry or other bodies. Consideration of the public interest is confined to the amelioration of death and injury in road accidents, either through accident avoidance, protection of occupants in the event of an accident or reduction in post-accident hazards such as protection against fire.

The NHTSA's budget provides for the sponsorship of advanced research in vehicle safety and this had led to a programme for Experimental Safety Vehicles (ESVs) intended to demonstrate the practicability of survival in accident situations more extreme than those which are taken as the norm for current vehicle designs, e.g. frontal impact into a 200-ton concrete block at 30 m.p.h. is today's standard but the ESVs are intended to withstand 50 m.p.h. impacts. Since kinetic energy is proportional to the square of the speed, the energy dissipation for the ESVs is $25/9 = 2{\cdot}78$ times the current standard. The barrier impact is equivalent to the impact between two similar cars with a closing speed of twice the barrier figure, i.e. 30 m.p.h. barrier equals 60 m.p.h. car-to-car, and 50 m.p.h. barrier equals 100 m.p.h. car-to-car, and the complete protection of occupants in such crashes is clearly very demanding in technical terms.

Much controversy in engineering and medical circles surrounds the human tolerance which may be assumed in order to derive standards for the maximum severity of the man/machine conflicts which occur in accidents. Yet a good law demands the specification of parameters and test methods in such a way that the results are repeatable with sufficient accuracy as to demonstrate achievement of a standard.

One way to avoid some of these difficulties is to require certain design features in a motor car which commonsense and a limited study of accidents indicates will be beneficial. A well-known example is the collapsible steering column intended to prevent impalement on the steering column shaft in a severe front impact. This was a frequent cause of American road deaths until new collapsible designs were introduced. The design specification of the vehicle to ensure "collapsibility" of the column was based on a 5-in. limit to the rearward movement of the steering wheel in frontal barrier impact at 30 m.p.h.

American safety standards such as this have been largely design orientated, but it is the stated intention of the NHTSA to replace many existing standards by a much smaller number of performance standards. These will lay down injury criteria for car occupants in the event of an accident, leaving the car manufacturer greater freedom in the choice of designs to meet these criteria. The difficulty of defining injury criteria and test methods which are truly representative of the majority of human beings and accident situations is the rock on which this approach could founder.

Human beings vary enormously, not only in their physical characteristics but in their ability to withstand and sustain damage in accidents. Furthermore, some forms of injury may be less acceptable than others. For example, broken bones and less obvious scars are generally more acceptable than brain damage, facial disfigurement and loss of sight. Accidents also vary enormously, and it is only possible for legislators to select particular combinations of all reproducible test conditions representing a very narrow band of the human and accident characteristics which occur in real life. At the same time, unless such conditions are specified, it becomes impossible to demonstrate that a manufacturer has designed his product to meet the requirements of the law and we thus have a somewhat artificial situation in which car design is being channelled to meet certain rather specific safety criteria, which it is hoped will give a high degree of protection to the majority of people in the majority of accidents.

Another problem in testing the safety performance of a vehicle design is the impossibility of crashing the vehicle with human occupants. Dummy people have been developed which simulate human beings in some respects. Made of metal and plastics they can be instrumented to determine the forces to which they may be subjected in a simulated or real crash. Analysis of the electrical signals from the instruments is an established part of the many crash tests now done every day. These are generally analysed using computers and give reasonably accurate facts on the dummy's reactions in the crash. Unfortunately dummies which are robust enough to use in this way behave differently to humans. If they could be made fragile and representative of human characteristics of bone, muscle, brain, etc. their test lives would become impossibly short—often just one test. Practicable dummies are, therefore, not entirely satisfactory as representatives of humans and acceleration figures taken from instruments installed in them are as yet very poor indicators of human survivability, despite NHTSA insistence that limits to these figures measured in head, torso and femurs of dummies are to be used to determine compliance with the performance standards required for occupant protection in 1975 and later models.

Consideration of the mechanics of occupant response in frontal impacts coupled with the NHTSA's proposed injury criteria mentioned in the preceding paragraphs has led to much interest in the air bag as a device for cushioning occupants in accidents. Successful demonstrations have been given in which dummies have been so cushioned in 30 m.p.h. barrier impacts, the air bag inflating in the few milliseconds between initial impact with the barrier and the moment when the occupant begins to move forward within the car. Instruments have measured acceptable decelerations of the dummy and there is here an apparent solution to the 1975 requirements, having the principal merit that an occupant is protected without any special action on his part. However, it is expensive and for continued protection over the life of the car demands space-age standards of reliability as yet unachieved in space rocket performance.

It also has some unfortunate side effects which make many of those who are familiar with the system, reject the concept out of hand. Anyone who has witnessed an air bag test will know that the noise level of about 170 dBA inside the car would normally be considered unacceptable, as it is loud enough to deafen permanently a substantial proportion of people.

Furthermore, the air bag can only be effective in certain types of crashes. Although it is theoretically possible to devise variants of the basic idea which would give a broader spectrum of accident protection, in practice only very simple variants may be feasible, leaving such circumstances as child protection inadequately catered for and roll-overs, multi-crashes, side impacts and frontal barrier impacts at speeds below 18 m.p.h. with no occupant protection whatever, although many injuries occur under these conditions.

In addition to safety controls, another American government agency, the Environmental Protection Agency (EPA) is charged with responsibility for controlling pollution of the atmosphere. Mobile sources (vehicles) must meet emission standards covering unburnt hydrocarbons, carbon monoxide and, in future, nitrogen oxides. Increasing severity is demanded until in 1975/6 emissions should contain only one-tenth of the three pollutants

compared with the average figures in 1970/1. It has been said that vehicle exhausts meeting these standards would eject from their exhausts cleaner "air" than they consumed although with very little oxygen and a large quantity of carbon dioxide.

Without exception vehicle manufacturers and other bodies have protested that these standards go beyond what is possible in the state of technical and scientific knowledge today, at least in the ability to mass produce such engine systems and to ensure that they continue to maintain their emission control performance throughout their useful life. Despite this, Congress passed the Clean Air Act amendments which required these stringent standards and President Nixon signed the Bill which had been originally sponsored by Senator Muskie.

The only amelioration provided for is a 1 year suspension of this law for a manufacturer who can show that despite good faith efforts to develop complying systems, he has been unable to do so. Only in extreme cases was such suspension possible under the terms of the Act. At the time of writing, all American automobile companies and Sweden's Volvo have given notice of their intention to apply for this suspension period. The National Academy of Science has pronounced a verdict against the feasibility of these standards at least in the time-scale proposed and a major confrontation between the manufacturers and the representatives of the law is being mounted.

I should introduce here the concept of cost/benefit analysis for evaluating the effectiveness of a proposed measure. This is often suspect in the eyes of the public, since it requires a highly questionable quantification of human values. How does one put a price on loss of hearing, facial disfigurement, loss of life? Insurance societies and the courts frequently do this even though the concept requires an equation between man and money which is dependent more on a judgement of equity than on any universal valuation of human suffering.

In the case of atmospheric pollution from vehicles, direct harm is more difficult to detect. Los Angeles photo-chemical smog is unpleasant but probably not lethal unless it has long-term effects on respiratory organs. In other parts of the U.S.A. photo-chemical smog is less severe. In the rest of the world it is generally no problem, although a few cities (e.g. Tokyo and Madrid) sometimes have trouble. The valuation of reduction in smog formation is uncertain. The costs of controlling emissions which cause it are enormous. The cost/effectiveness sum is inconclusive and politicians' law-making is based on factors other than a quantifiable return. In the case of advanced safety and pollution controls there has been more than a hint that the science and technology which can put a man on the moon and bring him back alive will make these goals achievable. This may well be possible but only perhaps if cost and benefit are dissociated.

I should perhaps mention here that the NHTSA made the statement in October 1971 that "Approval of rulemaking plans is based on a careful analysis of safety payoff in terms of lives saved and reduction in injuries and on estimates of cost to the consumer". A public statement identifying the benefits and costs against each proposed rule has not yet been made.

It is interesting to note that a recent study sponsored by the White House concludes that the costs of meeting the 1975 safety and emission standards will far exceed the social benefits to be expected to result. The nett excess of cost over benefit for emission controls alone is put at $6300-million per annum from 1976 on. Included in these costs is an average price increase of $350 per car at 1971 prices. The 1976 safety requirements would add an estimated $523 to this for a total of $873 at 1971 prices.

3. THE REST OF THE WORLD

In Oceania and the other 154 countries outside the U.S.A. are the other half of the world's cars and it is undeniable that they are affected by American actions and particularly by legislative activity such as discussed in Section 2. The increasing cost of exported American cars, the diverging size requirements in a congested world and the growth of motor industries in most developed countries has been reducing exports from the U.S.A. to negligible levels when considered as a trading offset to the sales into America of cars produced in Europe and Japan. Canada is ignored here because of its comparability and its manufacturing exchange arrangements with the U.S.A.

As the world's biggest single market, the U.S.A. has been a prime target for world exporters who have been energetically ensuring that this market remains open to them despite the non-tariff barrier built from safety and emission legislation.

At the same time, compliance with American standards is not an automatic guarantee that all models can be made to meet similar standards if introduced elsewhere. This is because firms generally do not export their whole product range to America but confine the cost and work load of meeting special U.S.A. requirements to one or two of the most suitable models in their range.

Extensive research and development programmes to meet the American standards in a timely manner on those models which they do market gave foreign companies with the will or necessity to remain in the U.S.A. market a fund of special expertise which was to prove useful in adapting their products to meet the somewhat different standards required by other governments on their domestic cars. In the last 5 years particularly, governments everywhere have introduced safety laws, some of which are based on the American standards. In the majority of cases local laws have been written to suit local needs, with the result that their requirements differ from one another not only in detail but often in principle. Taken together with language and legal differences, the tremendous growth in world vehicle legislation has proved an enormous challenge to companies with a tradition of world exports.

The table shows the numbers of legislative standards or proposals yet to become effective on production vehicles (as at March 1972). These range from a simple requirement such as marking of controls to the complexity of the 1975/6 U.S. emission levels. It is immediately apparent from a study of the laws of each country that many requirements overlap in the sense that the same subjects appear repeatedly although the details vary in specific application and timing.

TABLE I. Imminent legislative standards affecting European vehicle manufacturers

Source	Enacted	Proposed	Total
U.S.A.	32	56	88
ECE	24	24	48
EEC	7	15	22
U.K.	4	2	6
Other	12	14	26
Total	79	111	190

Note:- Figures show position at end of March 1972.

Motor manufacturers are publicly on record since at least the 1968 Geneva Motor Show as pledged to work for and press wherever possible for harmonization of safety and emission laws. Only by this means can the multiplicity of requirements be condensed into a handleable number of rational production solutions.

There are some moves toward harmonisation, and Working Party 29 of the Economic Commission for Europe (a major division of the United Nations Organisation) has been instrumental in working persistently in this field with delegations from many countries of the world. The outcome of their work during the last 14 years has been to publish 26 ECE standards for safety related features of vehicles. Sponsored by at least two members each standard is presented to the United Nations for ratification, after which countries may adopt the provisions into their own law. Certificates granted for vehicle features conforming to the relevant ECE standard are mutually acceptable between countries which have adopted that standard and vehicles embodying those "E-marked" features can be sold in each of the countries without re-test of that feature.

Unfortunately, as the subjects became more complex the rate of progress to agree new standards slowed. Also the unenforceable nature of the process of adopting approved standards meant that many countries have not adopted them although they were signatories to the 1958 agreement which established the principles. Many countries became disenchanted with the slow rate of progress and made their own laws without waiting for the United Nations.

With the coming of the Common Market a new situation arose. Here was a small group of countries pledged to develop their economic wellbeing communally and including Italy, France and Germany—three of the top four European car producers. Small wonder that the Commission should soon turn to harmonized laws for vehicle safety within the Six. Directives to harmonize many aspects of safety were written based on ECE standards where these existed. Member States were given 18 months to introduce directives into their state law and are required to accept for sale any vehicle meeting these directives. Directives pre-empt state laws wherever there might be a conflict unless a specific waiver has been granted and this is a most unusual occurrence.

With the expected accession of the United Kingdom to the Common Market all four of the major European car manufacturing countries will be included and from July 1973 existing directives are due to be incorporated in British law (as well as that of other new members). Where there are differences the directives will pre-empt the British standards. This will create a group of 10 countries progressing towards common vehicle legislation not only in technical terms but also in interchangeability of approvals with about 90% of Europe's vehicle park under its control. This will undoubtedly be the strongest world force for harmonization of legislation and one can foresee its standards becoming the basis for United Nations work rather than the reverse, as has previously been the case.

4. ENSURING COMPLIANCE

The millenium is not here yet and nor are universal interchangeably acceptable safety and other laws between the Common Market countries, or for that matter almost any other countries in Europe, and this brings us to a consideration of the different methods of ensuring compliance with the law.

American safety laws require each company making or marketing vehicles to certify that each vehicle meets the applicable standards. The penalties for non-compliance when discovered are severe, including fines up to $1000 per violation (i.e. per vehicle) and the requirement that vehicles are corrected or replaced. Self-certification, as this system is called, places the onus on the manufacturer to see that his production quality controls maintain the standards required to meet the law, while his achievements are monitored by random testing of sample vehicles and inspection of records.

American emission controls impose a type approval system on top of this, in that sample vehicles must be submitted to the EPA having completed 50,000 miles road testing under specified test cycles. With the vehicles are records of their emission performance at intervals through the test which, together with records of engine emissions and emissions from other sample vehicles, constitute the evidence on which the test certificate is finally given. For a new model this procedure takes many months in addition to the time taken to develop the engineering and production solutions to the emission control package.

European compliance procedures generally require that a representative vehicle be submitted to the country's approval authority where the vehicle type is to be sold. This is tested for conformity with the laws of the country and, if accepted, receives a type approval certificate allowing such models to be sold. If E-marked features are embodied in that country's law, certificates showing conformance to the relevant ECE standard are accepted rather than testing that feature again.

The EEC Type Approval Directive states that compliance may be achieved and demonstrated in one member state, after which the vehicle approved may be sold in the others. Features not covered by E-marks or directives must still be approved by the country's approval authority and it is unfortunately true that some EEC countries have special features not required elsewhere. For example, France requires yellow headlamps and Italy has a special ground clearance requirement, while the U.K. will aim for lower noise levels and, of course, drives on the right.

It is expected to be some years before these differences are eliminated and in the meantime individual type approvals must be sought from each country.

The process of doing this is time-consuming and expensive, all costs being chargeable to the manufacturer, including visits by government officials to other countries. Coordination of type approvals with development programmes is difficult and the definitions of "types" lead into problems of designation and difference which are often settled by negotiation rather than logic. The price for not receiving approval is high—sales are banned—so manufacturers struggle to meet the demands in a timely and effective manner. The engineering or specification changes which are normally made on a continuing basis to improve the product, adapt to the market, or cope with supply problems, become difficult under this procedure since those which may affect an approved feature (and the whole car is approved as a "type") are required to be advised to the approving authority for a decision on re-submission for a new certificate of conformance.

In practice it is impossible for all changes of all models to be notified to all approval authorities and it becomes essential to exercise judgement as to those which are important. The increasing proportion of the vehicle affected by legislation is bringing this problem to the fore and some resolution will be required before long if vehicle production is not to grind to a halt through bureaucratic controls. The British industry has recommended self-certification à la Americaine, but European Type Approval is firmly entrenched and no doubt will have to be made to work effectively despite the increasing burden of legislative requirements.

5. CONCLUSION

It is apparent that world demand for safer and cleaner automobiles as voiced through our political representatives, has set in motion a juggernaut of legislation which will crush those manufacturers who do not keep ahead of it. The costs of meeting these laws will in the end be borne by the customer or by the public, as must the social and other costs of *not* improving the safety or emission standards. It is hoped that the cost/effectiveness judgements made in our names will prove to be valid since the costs involved are proving to be very high.

It is no longer possible for safety or pollution control features to be incorporated voluntarily by a manufacturer as a sales feature, since the whole field of possibilities is covered by government action with only a few chinks left for manufacturers' initiatives. The speed of action required by governments demands that first priority be assigned to meeting the law. This diverts technical expertise and investment from programmes intended to improve business competitiveness and can be seen to be leading to less frequent model changes and a reduction in the range of available consumer choice. At the same time an increasing proportion of car design is dictated by legal rather than customer standards and it seems inevitable that differences between makes and models will continue to diminish.

Conflicting views between countries on the specific requirements of an item of vehicle legislation make great difficulties for exports and constitute trade barriers often of greater significance than the tariff barriers which they seem to have replaced. Moves to harmonize regulations between countries are making some progress but it will be many years at today's rate before it is really effective. Even then, there may never be less than three world groupings of countries having harmonized laws. North America is clearly at the centre of one of these, with Europe and possibly the Common Market as the hub of a second group. Outside these is a third world of countries where traffic density and living standards are low. It is questionable whether these can afford the luxuries of safety and clean air as defined by the developed nations. It is certain that they will never achieve the density of car ownership of the developed nations, in which case they probably have no need of the legislative strait-jackets which our governments are fashioning.

MOBILE SOURCE AIR POLLUTION—WHO WON THE WAR?—I†
(Where are we?)

S. WILLIAM GOUSE, Jr

Carnegie Institute of Technology and School of Urban and Public Affairs, Carnegie-Mellon University, Pittsburgh, Pennsylvania 15213, U.S.A.

In Part I of this paper, we examine the history of mobile source air pollution control—particularly, control of undesirable emissions from automotive vehicles. It is concluded that the choice of the level of control required of the automotive vehicle in the mid 1970's was arbitrary and not in the best interest of the public. The costs will outweigh the benefits, especially when one considers the health effects of various pollutants and relative contribution of the more noxious air pollution constituents from other sources.

In Part II we examine how we arrived at this state of affairs and address the major questions that have not yet been resolved.

INTRODUCTION

It seems timely to focus on major issues that have been faced and have yet to be faced in our control of mobile source air pollution.

1. What are the substantive, technical, intellectual and analytical resources that must be brought to bear in order to solve the problem? How do these capabilities and resources differ from what is available?
2. What are the institutional, political, social and financial arrangements that are essential for success? How do these arrangements differ from the existing ones?
3. What impediments in whatever area must be removed before we can succeed?

In order to offer answers to the above questions, one must take a look at what has happened, where we are today and how far we have yet to go in the control of mobile source air pollution. Partial answers to the questions above are presented as conclusions and recommendations of this paper.

THE STAGE

On December 31, 1970 the Clean Air Amendments of 1970, PL 91-604 became law. It was the strongest

† Adapted from an oral presentation at the Joint ORSA/IEEE Meeting, October 28, 1971, Annaheim, California. Work supported by Carnegie-Mellon University.

piece of legislation relating to restoring and maintaining the quality of environment yet passed in the U.S.A. It gave many new powers to the Administrator of the Environmental Protectional Agency (EPA), forced cross licensing of air pollution control technology, and took final authority for air quality improvements away from local governments, if they did not respond within the time frame allowed in the law. In the case of mobile source air pollution, it set standards and warranty requirements.

With respect to automotive air pollution, the law calls for a 90% reduction in emissions from the levels of control achieved by 1970 automobiles; by January 1, 1975, in the case of carbon monoxide and unburned hydrocarbons; and by January 1, 1976, in the case of oxides of nitrogen. Details of the test procedure and later determination of the 1970 levels were to be left to the Administrator of EPA and have since been established.[1]

Since there were no Federal controls on oxides of nitrogen, the law calls for only a 90% reduction from 1970. However, in the case of unburned hydrocarbons and carbon monoxide, substantial reductions from precontrol vehicles had already taken place. So the 90% reduction really means an order of 98% (Figure 1)[2] or more reduction from precontrol vehicles; a strigent and not cost effective level of control when compared to other sources of the same constituents.

Information available from EPA[3-5] plus recent estimates of fuel penalties associated with oxides of nitrogen control in mobile sources indicates that by the mid seventies we will be paying the order of

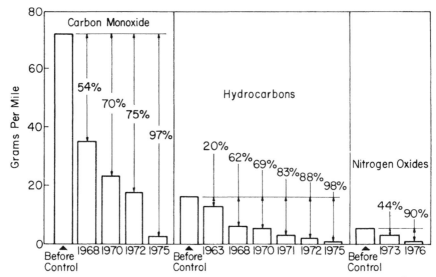

FIGURE 1   Accomplished reductions, and probable future reduction in vehicle emissions.[2]
Based on Federal Standards (California reductions generally accomplished 1–2 years earlier, except
for 1975–76).

$1000 per ton of oxide of nitrogen emission reduction from mobile sources compared to the order of $50 per ton[†] of oxide of nitrogen reduction from stationary power plants; assuming a 90% reduction from both types of sources. This is clearly not a cost effective approach, especially since, in the case of oxides of nitrogen, mobile sources are not generally the major source.

How did we get to this situation where standards were set in law by an overwhelming vote in the US House and Senate[6-8] in the face of claims that it was not technically possible and where the law itself indicates that technological feasibility and economics are not to be considered in implementation? Why was there rejoicing in this final blow to the automotive industry?

In order to understand where we are today and how we got there, one has to examine some of the history of the control of mobile source air pollution.

HISTORICAL ASPECTS OF MOBILE SOURCE AIR POLLUTION

The first serious attempts to deal with mobile source air pollution in the U.S. occurred around the turn

of the century. A recent article in the American Heritage Magazine[9] is an excellent summary of the state of affairs at the time. The problem was the horse. The average horse produces approximately 22 lb of solid waste and 1 gal of urine a day.[‡] Writers in popular and scientific periodicals were demanding "the banishment of horses from American cities". One authority wrote in 1908 that the 120,000 horses in New York City "were an economic burden and an affront to cleanliness and a terrible tax on human life". The solution to the problem of the horse, agreed the critics of that time, was the adoption (Figure 2)[10] of the "horseless carriage". In a city like Milwaukee in 1907, for instance, with a human population of 350,000 and a horse population of 12,500, the horse meant 133 tons of manure a day. Or as a health official in Rochester calculated in 1900 "15,000 horses in that city produced enough manure in a year to make a pile covering an acre of ground 175 ft high and breeding 16 billion flies". In addition, there was a serious abandoned dead horse problem not unlike abandoned auto problems. Owners of horses tended to leave the dead animals where they fell. They were even more difficult to trace than today's registered abandoned automobiles.

---

† The absolute value of these numbers is subject to argument. The relative magnitudes are firm.

‡ At this writing I have not been able to obtain data on the gaseous effluents of horses.

Operation of auto more economical than horse and carriage.

Offers a means of cheap mass transit.

Use of trucks in business enterprises (transport of goods) should increase net return on investment as compared to horse-drawn vehicle.

Should provide increased storage space because auto takes up less room than horse and wagon.

Fast response/little upkeep/fast refueling/more reliable than a horse.

Should permit rapidly available medical and other emergency help

Should enable farmers to market produce over a much wider geographical area.

Safer than horse because of better control and braking.

Offers superior maneuverability and imperviousness to weather conditions and fatigue.

Should relieve traffic congestion on city streets.

Would rid cities of unsightly and unsanitary conditions attributable to the horse.

Would improve health conditions, resulting in a reduction of infectious diseases and cases of diarrhea; would provde better ventilation for passengers.

Would restore frayed nerves and reduce nervous diseases from hectic pace of life in urban-industrial society.

Should reduce city noise.

Should extend and enhance rural life.

Would permit movement to suburbia.

Expected to halt movement of population from rural to urban areas.

Source: Charles G. Burck, "The Coming of Automobile Consciousness III" *Fortune* pp. 160–161. (January, 1971).

FIGURE 2 Selected turn-of-the-century expectations for the automobile.[10]

OTHER HISTORY

Control of air pollution predates even that of the problems presented by the horse. Reference 11 deals with the history of air pollution control in Allegheny County, Pennsylvania and Ref. 12 is concerned with various legal aspects of the history of air pollution control. There was much concern about air pollution resulting from locomotives in the 1800's. However, the first serious attempt to control air pollution in the United States were the Smoke Control Regulations implemented in the late 1940's (a major impact was the conversion of home heating from coal to gas or oil) in Pittsburgh, Pennsylvania. This was followed by work centered around automotive related problems, first in California and then nationally. References 13–15 present additional historical information on the control of air pollution.

CONTRIBUTIONS OF MAN

In order to put the levels of control presently legislated in the Clean Air Amendments of 1970 in some perspective, it's useful to examine the amount of pollution caused by man himself. The Bioastronautics Data Book, NASA Special Publication 3006[16] contains much information on the waste production of man. Accepted ranges of values are:

The average urine output is the order of 1·25 liter per man-day.

The wet weight of feces per man-day ranges from 0·25 to 0·81 lb.

The volume of flatus produced per day ranges from 0·8 to 5 liters.

In addition, there are other wastes such as sweat, hair growth, nails, dry skin flakes, miscellaneous skin secretions, etc.

In looking at the wastes of a human being, one must also be aware that agricultural wastes exist because of food production to support the human.[17,18] The BOD sewer load of our farm animals is equivalent to five times that of present population. There are also contributions from pesticides, herbicides, chemical fertilizers, erosion and other wastes associated with farming. Any policy that leads to significant increase in food consumption, population, or intensity of food production will have enormous adverse environmental impacts.

AUTOMOTIVE VEHICLE'S SHARE OF AIR POLLUTION

The contribution of the automotive vehicle to our total air pollution problem may be viewed on a tonage basis; on a short term, peak exposure, health effects basis; on a long term, low level exposure, health effects basis; on a national level; and on a local level. Figure 3†,[2] a Table based on data assembled by the National Air Pollution Control Association in 1968,[19] indicates the major constituents of air pollution on a weight basis and the major sources on a percent by weight basis independent of constituent. It shows that transportation provides 42%, on a weight basis, of all air pollutants. More recent data shows changes in the transportation contribution—generally, upward.

† The reader should not be mislead by the precision implied by the number of significant figures.

Source	CO	Part.	SO$_x$	HC	NO$_x$	Total	Distribution (%)
Transportation	63·8	1·2	0·8	16·6	8·1	90·5	42·3
Fuel combustion in stationary sources	1·9	8·9	24·4	0·7	10·0	45·9	21·4
Industrial processes	9·7	7·5	7·3	4·6	0·2	29·3	13·7
Solid waste disposal	7·8	1·1	0·1	1·6	0·6	11·2	5·2
Miscellaneous†	16·9	9·6	0·6	8·5	1·7	37·3	17·4
Total	100·1	28·3	33·2	32·0	20·6	214·2	100·0

†Primary forest fires, agricultural burning, coal waste fires.
Source: NAPCA Inventory of Air Pollutant Emissions, 1970.[19]

FIGURE 3   Estimated nationwide emission (millions of tons per year, 1968).[2]

Lave and Seskin,[20] looked at the health effects of long term exposures to various levels of air pollutants; in particular, oxides of sulfur and particulates. They indicated that the national economic impact of automotive pollution on health accounts for 25%. A more recent study by EPA itself,[21] indicates that the percentage is even smaller—the order of 9–10%.

Figure 4 shows the health effect of the 1968 contribution of automotive vehicles to the total air pollution inventory, using the Air Quality Standards of the California Air Resources Board (as of September 1969) as a measure of the deleterious effects of various constituents; an arbitrary ranking, but still one which tends to put the contribution of the automotive vehicle on a national basis in better perspective. However, in some urban areas, the impact of the auto on health is considerably greater.

The Clean Air Amendments of 1970 deal with stationary sources as well as mobile sources. However, the Legislative pressure in terms of 90% reduction for automobiles is not in keeping with the actual problem.

REQUIRED LEVEL OF CONTROL

Figure 5[22]†‡ illustrates the difference between what had been proposed by HEW/EHS/NAPCA§ for 1975 and what is called for in the Clean Air Amend-

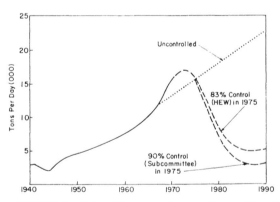

FIGURE 5   Nationwide automotive emissions, oxides of nitrogen.[22]

ments of 1970 for oxides of nitrogen. Figure 6[23] shows the difference between the Clean Air Amendments of 1970 and what HEW had planned for 1975 and 1980, assuming the 1980 research goals had become standards.

† Based on test procedures in effect at that time.
‡ Such curves have been verified by other investigators.
§ Before EPA.

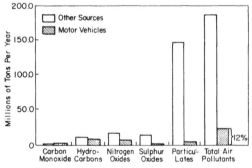

FIGURE 4   Motor vehicle contribution to U.S. air pollution based on harmfulness of pollution.
Based on desirable air quality standards per California Air Resources Board. September 1969.

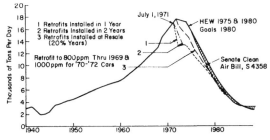

Source: S. William Gouse, Jr., "Retrofitting Pre-Emission Control Automotive Vehicle," Office of Science and Technology Memorandum, Oct. 23, 1970

FIGURE 6   Nationwide automotive oxides of nitrogen emissions.[23]

Figure 6 indicates that the difference, compared to other sources of air pollutants (Figure 7), is not significant even in California. Why then this more stringent control by Congress than proposed by the then NAPCA, now EPA? Is this small but expensive difference in level of control significant even if achievable in mass production? No. Not on the basis of present knowledge.

In order to understand what took place in the U.S. Congress in the late summer and fall of 1970 with respect to automotive emissions, one must go back to see how the automotive industry responded to public pressure concerning this problem Appendix A).

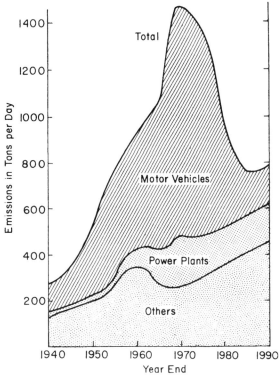

FIGURE 7   Estimates of oxides of nitrogen emissions in the South Coast Basin.[24]

APPENDIX A†

A Brief Chronology of Events in the Control of Air Pollution from automobiles

1. In 1947 the Los Angeles County Air Pollution Control District was formed.
2. In 1949 California Institute of Technology Biochemistry Professor Arie J. Haagen-smit, discovered the nature and causes of photochemical smog. His experiments revealed that the reaction of sunlight with unburned hydrocarbons and the oxides of nitrogen produced irritants.
3. In the period 1948–1957 the Los Angeles City Air Pollution Control District conducted an intensive program to control emissions of pollutants from industrial and domestic sources.
4. The Bureau of Air Sanitation was established in the State Department of Public Health by the California Legislature.
5. In July 1955 Congress passed Public Law 84–159 which was to provide research and technical assistance relating to

air pollution control. This activity was located in the Department of Health, Education and Welfare. Primary responsibility for air pollution control remained with the state and local governments.

6. In 1959 the Department of Public Health in the State of California adopted statewide ambient air quality standards and motor vehicle emission standards.
7. In February 1960 Governor E. O. Brown in California called a special session of the Legislature to consider air pollution legislation.
8. In July 1960, California Assembly Bill 17 was signed into law and the State Motor Vehicle Pollution Control Board was organized.
9. In 1960, the Division of Air Pollution and the U.S. Department of Health, Education and Welfare was established. Public Law 86–493 directed the Surgeon General to study effects of vehicle emissions on public health.
10. In 1961, California adopted test procedures and criteria for crankcase devices and exhaust controls. The automobile industry announced it would voluntarily install crankcase devices, nationwide, in all 1963 model cars.
11. In 1962 in the state of California, two crankcase emission control devices for used cars were approved.
12. In 1962, Congress made vehicle pollution studies the permanent responsibility of the Surgeon General.
13. In July 1963, California Senate Bill 325 established enforcement and compliance rules for control devices. The

_____

† Appendix A is a brief chronology of events in the control of automotive air pollution. Examination of this listing gives one a sense of the increasing pressure for control of automotive pollution and the rapid rate of increase of governmental regulations. Much of this was supplied by the California Air Resources Board.

first application for exhaust control recieved from a major auto manufacturer was from Chrysler Corporation.

14. In December 1963, the First Clean Air Act, Public Law 88-206, was passed. It authorized grants to go directly to Air Pollution Control Agencies on a matching basis. The matching was two for one for local work and three for one for regional work. It gave limited legal regulatory authority to Federal government and directed the Department of HEW to issue biannual reports to Congress on progress in abating automotive emissions.

15. In June 1964, California approved four muffler type exhaust control systems for factory installation in new cars and approved one for installation in used vehicles. In August 1964 the Big Four automobile manufacturers announced that they would equip 1966 models in California with approved exhaust emission controls.

16. In 1964 Public Law 88-515 required the general services of administration to use the California standards of 1964 for Federal vehicles being purchased.

17. In November 1964, the first exhaust control system from a major automobile manufacturer, Chrysler Corporation, was approved by California.

18. In October 1965, Public Law 89-272 was passed. Title II of the Clean Air Act was amended to authorize national regulation of air pollution from new motor vehicles. The first standards were to be applied to 1968 model cars and light trucks with the same standards to prevail in 1969. Sample vehicles could be submitted to HEW by manufacturers on a voluntary basis for testing and certification. That sample was to cover all vehicles of that production run.

19. On March 30, 1966, the Federal government published National Emission Standards for the 1968 and 1969 model year.

20. In June 1966, the State Board of Public Health, State of California adopted Diesel Odor Standards.

21. In August 1966, California approved the exhaust control systems for 1967 model cars.

22. In September 1966, the California Highway Patrol began random road side checks on the mechanical condition of vehicles including smog device installations.

23. In May 1967, the first 1968 auto exhaust emission control system certificate of approval goes to the Toyota Motor Company for sale of vehicles in California.

24. In August 1967, the California Senate Bill 490 created California Air Resources Board.

25. In November 1967, the Air Quality Act of 1967, Public Law 90-148, was passed. It provided for new mechanisms for attacking air pollution problems on a regional basis, provided for the establishment of a Presidential Air Quality Advisory Board, gave HEW authority to require registration of fuel additives and made possible the exception of the State of California from national standards.

26. In November 1967, the Dingell Amendment of the Federal Air Quality Act of 1967, which could have prevented California from imposing stricter vehicle emission standards for 1970 models, was defeated in the U.S. House of Representatives by a vote of 322 to 0.

27. In June 1968, Federal Hearings were held in San Francisco to determine whether or not the U.S. Department of Health, Education and Welfare should wave for California application of revisions of the Air Quality Act which provides a control that motor vehicles be exclusively province of the Federal government.

28. February 8, 1968, the first meeting of the California State Air Resources Board was held in California.

29. March 1968, Dr. Arie J. Haagen-smit was appointed Chairman of the Air Resources Board in the State of California by Governor Reagan.

30. In April 1968, the California Air Resources Board voted to require evaporative loss controls beginning with the 1970 models on vehicles sold in the State of California.

31. On June 4, 1968, the Federal Standards for exhaust emissions, fuel evaporative emissions and smoke emissions applicable to 1970 and later vehicles and engines were announced.

32. In July 1968, Department of HEW Secretary Wilbert Cohen grants waver to California giving the state the right to enforce its own vehicle exhaust emission standards for 1969 and 1970 model cars and light duty trucks.

33. In August 1968, Governor Reagan signed California Assembly Bill 357, the Pure Air Act of 1968 establishing progressively stricter emission standards for 1970–1974 model vehicles on a 5 year basis. The state requested a new waver from HEW to enforce these stricter standards—the strictest set anywhere.

34. In September 1968, the California Air Resources Board moved its headquarters from Los Angeles to Sacramento. The motor vehicles Laboratory continued to operate in Los Angeles.

35. In November 1968, the California Air Resources Board divided California into 11 air basins according to topography and climate.

36. In March 1968, the Air Resources Board certified the first evaporative control (Ford Motor Company for its 1970 Maverick).

37. In May, 1969, Department of HEW's Secretary Robert Finch grants Federal waver requested by California to allow the state to enforce stricter motor vehicle standards than those in the other 49 states.

38. From September to November 1969, statewide ambient Air Quality Standards were adopted by the California Air Resources Board for oxidants including ozone, carbon monoxide, H_2S, NO_2, SO_2, visibility reducing particles and particulate matter. The Air Resources Board Technical Advisory Committee proposed adoption of vehicle emission levels lower than those established by the Pure Air Act of 1968. The new standards were to go into effect with the 1975 model cars.

39. On November 20, 1969, President Nixon attended a meeting of his Environmental Quality Council† at which the chief executives of the major auto companies were present. At this meeting, President Nixon, through the Department of HEW and the Department of Transportation, outlined the proposed 1975 Federal Automotive Emissions Standards and 1980 research goals and the proposed Federal program to develop alternative power plants for automotive vehicles. The impact of that meeting on the auto industry was probably very significant. Shortly thereafter in December 1969 and January 1970 the presidents of the major companies made speeches indicating company policy to produce as clean an automotive vehicle as possible.

40. In January 1970, the Air Resources Board adopted new low emission standards to go into effect with the 1975 models. These were unburned hydrocarbons 0·5 g/mile, carbon monoxide 12 g/mile, oxides of nitrogen 1 g/mile.

41. On February 10, 1970, Present Nixon issued his environmental message announcing a number of major

---

† Before establishment of Council on Environmental Quality.

programs relating to restoring the quality of the environment. In this program he announced the establishment of a program to develop alternative power plants for automotive vehicle propulsion.

42. On March 19, 1970, the Office of Science and Technology Ad Hoc Panel on Unconventional Vehicle Propulsion issued a report backing the Federal decision to launch a program to develop alternative power plants for automotive vehicle propulsion.

43. In March 1970, Governor Reagan convened a joint meeting of the Air Resources Board and its Technical Advisory Committee. The presidents of the automanufacturing companies and major oil refineries met and discussed the emissions problem. The results of this meeting were that the Air Resources Board recommended reduction of lead in fuels beginning January 1, 1971 with the gradual phasing out over a period of 6 years. Starting on January 1, 1977, no motor vehicles' fuel sold in California could contain lead additives.

44. May 15, 1970, General Motors Corporation press release on retrofit kits to be installed on precontrolled vehicles beginning with program in Pheonix, Arizona.

45. On May 19, 1970, the President called for a tax on lead additives to automotive vehicle fuels as a means for reducing emissions of this substance from automotive vehicles.

46. On July 15, 1970, the Federal Register published proposed rule making and regulations for the control of air pollution from new motor vehicles and engines as well as a new sampling in analytical systems for measuring automotive exhaust emissions.

47. On December 31, 1970, Public Law 91-604 the Clean Air Act Amendments of 1970 were signed into law.

48. On June 29, 1971, the Federal Register published new Environmental Protection Agency procedures for certifying low emission vehicles. Section 212 of the Clean Air Act Amendments of 1970 sets up the procurement policy.

49. On July 2, 1971, the Federal Register published a final 1975–1976 automotive emissions standards and revised the test procedures published earlier.

REFERENCES†

1. "EPA issues final 1975–76 auto pollution standards, revises test procedures" *Federal Register* (July 2, 1971).
2. "Personal correspondence with Bruce Simpson" *Ford Motor Company, Dearborn, Michigan* (June 11, 1971).
3. "Control techniques for CO, NO and HC emissions from mobile sources" *USDHEW/PHS/EHS/NAPCA, AP-66.* USGPO $1.25 (March 1970).
4. "Control techniques for NO emissions from stationary sources" *USDHEW/PHS/EHS/NAPCA, AP-67.* USGPO $1.00 (March 1970).
5. "The economics of clean air" *Report of the Administrator of the Environmental Protection Agency to Congress* (March 1971).
6. "National Air Quality Standards Act of 1970" *Report of the Committee on Public Works, U.S. Senate to Accompany S-4358* (September 15, 1970).
7. "Clean Air Amendments of 1970" *Conference Report* (December 17, 1970).
8. "Clean Air Amendments of 1970" *PL* 91-604 (December 31, 1970).

9. "Environmental quality—2nd annual report" *Council on Environmental Quality* (August 1971).
10. Willis E. Jacobsen, "A technology assessment methodology: a pilot study on automotive emissions control" *The MITRE Corporation, MTR* 6009, **2** (June 15, 1971).
11. John Grove, "Background of air pollution in Pittsburgh —the forces which motivated the program" *ASME Paper* 59 *PBC*-1, Pittsburgh Bicentennial conference of ASME (April 20, 1959).
12. A. J. Petit-Clair, Jr., "Air pollution control: political solution or equitable resolution" *Congressional Record E* 4634-40 (May 19, 1971).
13. Oscar S. Gray, *Environmental Law-Cases and Materials.* The Bureau of National Affairs, Inc., Washington, D.C. (1970).
14. H. F. Lund (editor), *Industrial Pollution Control Handbook.* McGraw-Hill Book Co. (1971).
15. S. Edelman, *The Law of Air Pollution Control.* Environmental Science Services, 24 Danbury Road, Wilton, Conn. (1971).
16. Paul Webb (editor), *Bioastronautics Data Book.* (NASA SP-3006, USGPO 2.25 (1964).
17. "Animal wastes" *Staff Report, National Industrial Pollution Control Council USDOC* (February 1971).
18. "Animal slaughtering and processing" *Subcouncil Report, National Industrial Pollution Control Council USDOC* (February (1971).
19. *Nationwide Inventory of Air Pollution Emissions* "1968". U.S. Department of Health, Education and Welfare (August 1970).
20. L. B. Lave and E. P. Seskin, "Air pollution and human health" *Science* **169**, No. 3847, 723-733 (August 31, 1970).
21. L. B. Barrett and T. E. Waddell, "The cost of air pollution damages: a status report" *Air Pollution Control Office, EPA, Research Triangle Park, North Carolina* (April 1971).
22. *Statement by L. A. Iococca.* Ford Motor Company, Dearborn, Michigan (September 9, 1970).
23. S. W. Gouse, Jr., "Retrofitting pre-emission control automotive vehicles" *Office of Science and Technology Memo* (October 23, 1970).
24. "Air pollution control in California 1970" 1970 *Annual Report of the California Air Resources Board, ARB* 1108, *14th Street, Sacramento, California, 95814* (January 1971).
25. "Motor vehicles, air pollution and health" *Surgeon General's (U.S.A.) Report to Congress* (1962).
26. "Control of air pollution from new motor vehicles and new motor vehicle engines" *Federal Register* **31**, No. 11 (March 30, 1966).
27. "Air pollution—1966" *Hearings before a subcommittee on Air and Water Pollution of the Committee on Public Works U.S. Senate* (June 7-9, 14, 15, 1966).
28. "Proceedings: The Third National Conference on Air Pollution" *U.S. Department of Health, Education and Welfare, Washington, D.C.* (December 12-14, 1966).
29. "Electric vehicles and other alternatives to the internal combustion engines" *Joint Hearings before the Committee on Commerce and the Subcommittee on Air and Water Pollution of the Committee on Public Works,* U.S. Senate (March 14-17 and April 10, 1967).
30. "Automobile steam engine and other external combustion engines" *Joint Hearings before the Committee on Commerce and the Subcommittee on Air and Water Pollution*

† This Reference list concerns both parts of the paper.

of the Committee on Public Works, U.S. Senate, Serial 90–82 (May 27–28, 1968).

31. "The automobile and air pollution—a program for progress" *A Report from the Panel on Electrically Powered Vehicles of the Commerce Technical Advisory Board*, U.S. Department of Commerce (December 1967).

32. "Standards for exhaust emissions, fuel evaporative emissions and smoke emissions, applicable to 1970 and later vehicles and engines" *Federal Register* **33**, No. 108 (June 4, 1968).

33. "Air quality criteria for particulate matter" *USDHEW/PHS/NAPCA*, AP-49. USGPO $1.75 (January 1969).

34. "Air quality criteria for sulfur oxides" *USDHEW/PHS/EHS/NAPCA*, AP-50. USGPO $1.50.

35. "Control techniques for particulate air pollutants" *USDHEW/PHS/EHS/NAPCA*, AP-51 (January 1969).

36. "Control techniques for sulfur oxide air pollutants" *USDHEW/PHS/EHS/NAPCA*, AP-52 (January 1969).

37. "The search for a low emission vehicle" *A Staff Report Prepared by the Committee on Commerce, U.S. Senate*. 91 Congress, 1st Session (1969).

38. "Consent decree" *U.S. versus AMA Inc.*, GMC, FM Co., C. Corp., AMC. (Filed September 11, 1969, Final Judgement October 29, 1969).

39. "Control of vehicle emissions after 1974" *Report to California Air Resources Board by the Technical Advisory Committee* (November 19, 1969).

40. Andrew Jamison, *The Steam Powered Automobile—An Answer to Air Pollution*. Indiana University Press (1970).

41. John C. Esposito, *et al.*, *Vanishing Air*. Grossman Publishers, New York (1970).

42. P. S. Myers, "Automobile emissions—a study in environmental benefits versus technological costs" *SAE Paper* 700182 (January 1970).

43. "Presidents environmental message" *Announcing AAPS Program* (February 10, 1970).

44. "Air pollution—1970" *Hearings before the subcommittee on Air and Water Pollution of the Committee on Public Works*, U.S. Senate (Part 1—March 16–18, 1970); (Part 2—March 19–23, 1970); (Part 3—March 24–25, 1970); (Part 4—March 26, April 1, 17, May 27, 1970) (March 1970).

45. "Air quality criteria for carbon monoxide" *USDHEW/PHS/EHS/NAPCA*, AP-62. USGPO $1.50 (March 1970).

46. "Air quality criteria for photochemical oxidants" *USDHEW/PHS/EHS/NAPCA*, AP-63 (March 1970).

47. "Air quality criteria for hydrocarbons" *USDHEW/PHS/EHS/NAPCA*, AP-64. USGPO $1.25 (March 1970).

48. "Control techniques for CO emissions from stationary sources" *USDHEW/PHS/EHS/NAPCA*, AP-65. USGPO $0.70 (March 1970).

49. "Control techniques for HC and organic solvent emissions from stationary sources" *USDHEW/PHS/EHS/NAPCA*, AP-68. USGPO $1.00 (March 1970).

50. *Report of the Ad Hoc Panel on Unconventional Vehicle Propulsion*. Office of Science and Technology, Washington, D.C. (March 19, 1970).

51. Donald K. Kummerfeld, *Federal Policy on Auto Air Pollution Control*. Center for Political Research, Washington, D.C. (April 13, 1970).

52. B. H. Eccleston and R. W. Hurn, *Comparative Emissions from Some Leaded and Prototype Lead-free Automobile Fuels*. Bureau of Mines Report 7390 (May 1970).

53. W. P. Lear and K. L. Nall, "Vehicle air pollution—the problem and its solution" *SAE Paper* 710272 (May 12 1970).

54. "Control of air pollution from new motor vehicles and new motor vehicle engines" *Federal Register*, **35**, No. 136, Part II (July 15, 1970).

55. "Environmental quality" *The First Annual Report of the Council on Environmental Quality* (August 1970).

56. *A Survey of Propulsion Systems for low emission Urban Vehicles*. The MITRE Corporation, Report M70-45 (September 1970).

57. Memo for Dr. E. E. David, Jr. from Dr. S. W. Gouse, Jr., "National Air Quality Standards Act of 1970" S-4358 (October 9, 1970).

58. "IIEC: A co-operative research program for automotive emission control" *SAE Congress*, SP-361, Detroit, Michigan (January 11–15, 1971).

59. "Medical aspects of air pollution" *SAE Continuing Education Course* (January 14, 1971).

60. "Air quality criteria for NO" *EPA*, AP-84. USGPO $1.50 (January 1971).

61. John N. Pattison, "New federal driving cycle for emission tests" *APCA Journal* **21**, No. 2, 88–89 (1971).

62. "Personal correspondence with Sidney A. Mandel" *Information Officer, California Air Resources Board*, *ARB Chronology and Accomplishments*, Sacramento, California (April 26, 1971).

63. "Environmental conservation" *The Oil and Gas Industries*, Vol. I, *A Summary, National Petroleum Council* (June 1971).

64. Jon H. Heuss, George J. Nebel and Joseph M. Colucci, "National air quality standards for automotive pollutants: a critical review" *GMR*-1108, *Presented at the 64th Annual Meeting of the Air Pollution Control Association, Atlantic City, New Jersey* (June 27–July 2, 1971).

65. "Outgoing APCA president calls air standards program 'Almost ludicrous' " *Air/Water Pollution Report* (July 5, 1971).

66. E. Starkman, "The chances for a clean car" *Astronautics and Aeronautics* 68–75 (August 1971).

67. Joel Tarr, "The horse—polluter of the American city" *American Heritage Magazine* (October 1971).

68. "Heavy-duty engines proposed 1973 emission standards" *Federal Register* **36**, No. 193, 19400–19406 (October 5, 1971).

69. "Conference on low pollution power systems development" *Eindhoven, The Netherlands* (February 23–25, 1971) *Committee on the Challenges of Modern Society, NATO*.

70. "A factual record of correspondence between Kenneth Hahn, Los Angeles County Supervisor and the presidents of General Motors, Ford and Chrysler regarding the automobile industry's obligation to meet its rightful responsibility in controlling air pollution from automobiles—February 1953—January 1967" Los Angeles County (January 1967).

71. H. J. Hall and W. Bartok, "NO$_x$ control from stationary sources" *Environmental Science and Technology* **5**, No. 4, 320–326 (1971).

72. O. S. Barth [J. C. Romanovsky, E. A. Schuck and N. P. Cernansky]† "Federal motor vehicle emission goals for CO, HC and NO$_x$, based on desired air quality levels" *APCA Journal* **20**, No. 8, 519–523 (August 1970).

---

† Did not appear on APCA Paper.

73. B. H. Eccleston and R. W. Hurn, "Exhaust emissions from small, utility, internal combustion engines" *Work in Progress, Bartlesville Petroleum Research Center, Bureau of Mines.*

74. "Automotive fuels and air pollution" *Report of the Department of Commerce, Technical Advisory Board, Panel on Automotive Fuels and Air Pollution* (March 1971).

75. H. J. Wilmette and R. T. Van Dervees, *Report on the determination of mass emissions from two-cycle engine operated vehicles.* Olson Laboratories, Inc., Dearborn, Michigan, PB-194145 (January 23, 1970).

76. W. F. Marshall and R. D. Fleming, *Diesel Emissions Reinventoried.* Bureau of Mines, RI 7530 (July 1971).

77. F. L. Voelz, "Survey of automotive exhaust emissions and specific engine malfunctions—Summer 1970" *Atlantic Richfield Company, Harvew, Illinois, Presented to ACAAPS,* (October 22, 1971).

78. "Atlantic Richfield Company clean air caravan results—nationwide summary" *ARCO, Harvey Technical Center, Harvey, Illinois.*

MOBILE SOURCE AIR POLLUTION—WHO WON THE WAR?—II
(How did we get there and where do we go?)

S. WILLIAM GOUSE, JR.

Carnegie Institute of Technology and School of Urban and Public Affairs, Carnegie-Mellon University, Pittsburgh, Pennsylvania 15213, U.S.A.

Based on the history presented in Part I, we can conclude that the U.S. Congress behaved unwisely when it came to setting the Automotive Emissions Standards. This behaviour was rationalized, to a certain extent, by the past performance of the automotive industry but was not in the best interest of the public. Without dealing with the problem of vehicles in the hands of owners and other sources of air pollution, the high acquisition and operating costs to be paid by purchasers of new automobiles in the mid 1970's will not lead to commensurate improvements in air quality.

INTRODUCTION

In Part I of this paper we set the stage for the mobile source air pollution problem, discussed some historical aspects of mobile source air pollution control, examined the contribution of man himself, the automotive share of air pollution on a weight and health effects basis, and some remarks on the required level of control. In this part of the paper, we examined how certain information has been misused in the setting of automotive standards, discussed relative control of internal combustion engines for various applications and the problem of keeping automotive vehicles in the hands of public performing at or close to the levels of performance when new.

PEOPLE VERSUS DETROIT

An excellent place to begin is the correspondence that took place[70]‡ between Kenneth Hahn, Los Angeles County Supervisor, and the Presidents of General Motors, Ford and Chrysler regarding the automobile industry's obligation in controlling air pollution from automobiles. This collection of correspondence has been published by the Los Angeles County Board of Supervisors and begins with a letter from Mr. Hahn to the Secretary of Commerce of the United States dated March 25, 1966.

‡ A complete Reference List appeared in Part I of this paper.

In addition, one should examine the record of public hearings held in the State of California relating to automotive air pollution and testimony which has been made before Congressional Committees in Washington.

This public record leaves one with an unmistakable impression that the automobile industry was dragging its feet. Whether this was true, cannot be determined from the correspondence but the impression was created and still exists. This testimony and correspondence plus the industries' response to legislation, i.e. producing the devices it had shortly before called impossible or unknown, has led to a credibility gap of enormous proportions.

There are a number of reasons for this industry response—some understandable, others justifiable by our economic system. In the early days of the confrontation, industry was as ignorant of the details of automotive air pollution as were the regulators. Industry also went through a period of virtue shock in which it reacted to the black hat which was being put on it and could see only the economic good that it had for the country. There are legal problems related to industry cooperating or agreeing to do things as well as public relations problems in terms of the kind of testimony the industry gives. Often, in testimony, when it said it didn't know how to do things, it meant that it didn't know how that day, not a year later. Many of industries' complaints about inability to respond have to do with the lead time associated with aspects of its business. It has never succeeded in making clear to the public what is involved in producing millions of something each year.

Finally, the issues became too political. It was no longer possible to make use of analytical results as to what were reasonable levels of control, especially in the light of the instructions in the legislative history of the Clean Air Amendments of 1970 which indicated that in setting standards, EPA should take the most conservative approach possible, i.e. any observable adverse effect on human beings. No distinction was made between reversible and irreversible effects, nor was there any discussion in the legislative history of the amendments as to what might be an acceptable social cost. Clearly, it is impossible to protect the entire population from adverse effect no matter what the level of control. Even natural backgrounds level will give some people problems.

If this only happened in one industry, we could probably tolerate the process. However, similar plays are being acted out with other industries. It seems that as a whole, the country has learned little from its attempts to bring mobile source air pollution under control.

MISUSE OF DATA

Environmental enthusiasts tend to misuse data as much as others in this game. One has often heard that the automobile causes 90% of the pollution, 60% of the pollution or 80% of the pollution. The government's own data indicates that on a tonage basis in 1969, transportation as a whole, did not contribute more than 50% of the national air pollution. However, certain constituents such as carbon monoxide were primarily from the automotive vehicle. In addition, in certain areas of the country such as the Los Angeles Basin, the automobile does produce the order of 90% of the air pollution.

Figure 1[71] is typical. While not untrue, the Figure presents a misleading set of information. It shows NO_x emissions increasing with time for all sources. The Figure clearly indicates potential cumulative NO_x emissions. But if one showed such a curve indicating what standards are already in effect and promulgated, then one gets a different impression. For example, Figure 7[24] (in Part I of this paper) estimates the NO_x emissions in the South Coast Basin of California in the Los Angeles area. This Figure clearly shows the motor vehicle contribution

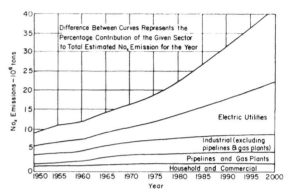

FIGURE 1 Potential cumulative NO_x emissions in the U.S. (medium economic trends).[2]

is under control with the emissions standards then promulgated to go into effect.†

The government has also done this in the past. Figure 2[19] taken from a report published August 1970 shows emissions increasing "for carbon monoxide in 1980 at a time when" 75 standards were in the final stages of rule making and would have become standard had the Clean Air Amendments of 1970 not passed.

People look at such data (Figure 2) and observe that things are only going to improve temporarily. They do not look behind the curves to see what levels of control are already in the works; though not in effect. Congress and the scientific community are both to be blamed—the scientific community

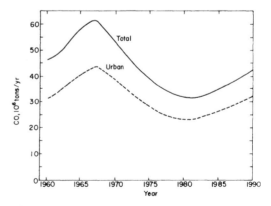

FIGURE 2 Carbon monoxide, emissions estimate based on present legislative standards.[4]

† This assumes an effective inspection and enforcement system to keep vehicles properly maintained.

for not carefully enough indicating what their results mean and the Congress for misusing results even when they have caveats in them. The committee report on the Legislative history of the Clean Air Act Amendments bears this out as does the Congressional record during the time of committee debate on the Clean Air Amendments. A paper by Lave and Seskin[20] on economic impact of air pollution deals mainly with sulfur dioxide and particulates, and yet had a major impact on the committee's deliberation on automotive air pollution. Most of the automobile's effluents were completely unrelated, at that time, to the work presented in that paper.

The level of control in the Clean Air Act Amendments of 1970 was strongly based on Barth's, et al.[72] work, even though it is indicated that they were presenting a technique rather than a set of emission standards.

The ambient air quality standards that have recently been promulgated by the Environmental Protection Agency are under considerable attack. They are based on the premise that no one shall experience an adverse health effect. Clearly, this is impossible. Evidence is beginning to accumulate that indicates that in some areas of the world natural background levels exceed the recent ambient air quality standards. In addition, one must distinguish between reversible and irreversible health effects and the fact that there will always be some segment of the population susceptible to even extremely low concentrations of air pollutants. The solution here is not to protect everyone, everywhere, but to provide a mechanism for a more cost effective control strategy than now in effect, for more stringent local standards, and for relocating some people and some sources.

The present level of expenditure on research into the health effects of long term, low level exposure to various constituents of air pollution is grossly inadequate. The amount of public expenditure that results from too tight an emission standard is way out of proportion to the present level of public investment. On the other hand, if for certain constituents extremely tight standards are required, it is best to do the research as soon as possible in order to establish these levels immediately. If extremely tight levels of control, perhaps tighter than already promulgated are required, we should know these soon, because the kind of problem that we face in meeting these standards and protecting public health have not been fully appreciated by those setting the standards or by the Committee

in Congress to which the standards setters are responsible.

RELATIVE CONTROL OF INTERNAL COMBUSTION ENGINES FOR VARIOUS APPLICATIONS

The law as now written applies stringent emission controls to gasoline powered automotive vehicles and light duty trucks. Heavy duty diesel powered trucks and buses are not stringently controlled† nor are evaporating losses that take place in the filling of automotive fuel tanks or in the transfer of fuels between refineries and automobiles. Nor are many other kinds of internal combustion engines used in motor cycles, lawn mowers, boating, stationary power units for construction work as well as power generation, snow mobiles, all terrain vehicles, chain saws, etc. under control.

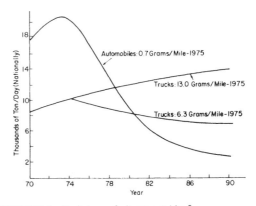

FIGURE 3   Emissions of nitrogen oxides.[8]

At one time, non automotive engines constituted a small contribution of air pollution compared to the uncontrolled automotive vehicle. This is no longer true. Existing legislation calls for 90–98% control of automotive emissions. This will make the sources of air pollution from other uncontrolled engines exceed those of automotive engines. Figure 3[74] shows what one can expect in terms of oxide of nitrogen emissions from heavy duty trucks if present California Standards were imposed nationally. It is clear that these are not stringent enough in comparison to the automotive

† Standards for control of emissions from heavy duty engines have been proposed.[73]

vehicle. Figure 4[8] shows the amount of unburned hydrocarbon associated with servicing vehicles.

Recent works under way in various laboratories indicates that small[7,75,76] 2- and 4-cycle engines are very dirty compared to the automotive engines now being produced, and, the rate at which the market for these uncontrolled vehicles is increasing is very significant. They will soon be major sources of hydrocarbon, carbon monoxide and oxides of nitrogen. A careful benefits/costs analysis will have to be made to understand the levels of control required for all these sources. In addition, the lead times required will be very significant. Many of these other power plants are made by small manufacturers. In addition, small power plants do not have much unused capacity for the operation of emission control equipment.

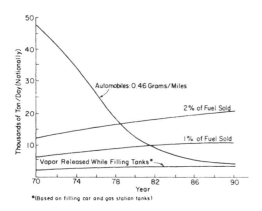

FIGURE 4   Emissions of hydrocarbons.[8]

RETROFITTING AND INSPECTION

Major questions that have not yet been faced in the control of mobile source air pollution have to do with existing, uncontrolled vehicles and the maintenance of all vehicles in operation.† Data is now available[77,78] that indicates that precontrol

† With respect to the Clean Air Amendments of 1970, issues relating to assembly line inspection, the kind of maintenance allowed during certification and whether or not the average vehicle or every vehicle must meet the Emission Standards, have not yet been resolved. These issues have enormous cost impact and greatly affect the design and manufacturing process. If we go to extensive inspection systems once vehicles are in the hands of owners, then we must have some way of knowing what the emission performance was when it left the plant in order to know whether the vehicle passes or fails an inspection test.

vehicles, especially those in the state of poor maintenance, are extremely dirty compared to the levels of control called for in the middle of this decade. Vehicles out of tune or with misfiring spark plugs can be 50–100 times as dirty as new vehicles. Even though they represent a small share of the total vehicle population and are not driven as much as our newer vehicles, they still make a substantial contribution to mobile source air pollution.

The Clean Air Amendments of 1970 specify that there shall be warranty requirements of 5 years or 50,000 miles; details of which are yet to be determined. The law allows that the manufacturer can require reasonable maintenance requirements on the part of the owner. Experience with past warranty requirements by the industry indicates that between 10% and 30% of owners comply with maintenance requirements.

Unless one institutes an inspection system, it is reasonably clear that many operators will not comply with emission control system maintenance requirements, and therefore, that their vehicles will shortly after purchase be operating with degraded emission control performance. Often, the vehicle will serve adequately as a transportation device even though it is emitting more than it is designed to emit.

With the level of control now being called for—the order of 98% on hydrocarbon and carbon monoxides, 90% on oxides of nitrogen, it is essential that inspection systems and retrofit programs be initiated if the level of expenditure for vehicles produced in 1975 and beyond are to have any meaning at all in terms of control of mobile source air pollution. Present estimates indicate the order of $200–250 per car to meet the 1975 standards and upwards from there to meet the 1976 Oxides of Nitrogen Standards. There can be no justification for increasing the initial purchase price of vehicles this amount if programs are not initiated to make certain that a vehicle is properly maintained to keep its emission characteristics in a desirable range.

CONCLUSIONS AND RECOMMENDATIONS

Let us return to the questions in the Introduction and reply in terms of the material presented here and cited in the References (see Part I). One must keep in mind that it would not be possible to present all of the information in detail.

(1) What are the substantive technical, intellectual and analytical resources that must be brought to bear in order to solve the problem? How do these capabilities and resources differ from what we have available?

1. One must recognize that air is being polluted. This has happened.

2. One must understand the effects of various types and levels of air pollutants. There are regional effects, health effects and physical property damage effects. We have some understanding of short term high levels of exposure and how these effect health. Our understanding of long term effects of low level of exposure is poor.

While it is true that one cannot wait until all the information is in hand before imposing emission standards, one can be more relaxed with respect to mobile source air pollution than is now the case. In many areas, even zero emissions from automotive vehicles would make little impact on total air quality.

3. One must know the distribution of sources of various air pollutants and their magnitudes. Instrument stations are being put on line in various air quality control regions around the country, and a data base is being established. In some regions data goes back a considerable number of years. Unfortunately, techniques or instrumentation often used are now unaccepted.

4. One must be aware of the synergestic effects that may result from interaction of various air pollutants. In this area, our knowledge is very incomplete.

5. One must understand the differences between mobile and stationary sources, their distribution and magnitudes. In some regions of the country that have received intensive study, this information is available. It is known that in some places, the mobile sources are the controlling sources; while in other areas of the country, stationary sources control air quality.

6. One must have a good knowledge of natural sources of air pollution as opposed to man-made sources. Our knowledge in this area is marginal. Extensive experimental efforts must be mounted.

7. One must understand the effects of various possible control processes. For example, in the control of hydrocarbon and carbon monoxide emissions from automotive vehicles, one increases the emission of oxides of nitrogen. Or, for example, in the control of the air pollution emissions of various industrial processes one creates intolerable solid waste or water pollution problems. We are learning more about such possible counter productive effects which result from the imposition of levels of control of some undesirable emissions without fully understanding the problem. Again, difficulty is not allowing enough flexibility in our standards and enforcement systems to permit readjustment of standards when such undesirable situations are encountered.

8. One must understand the life cycle of various air pollutants and the time constants associated with them. This is an area of activity which needs much attention.

9. One must have good data on the meteorology of various regions and understand meteorological phenomenon such as inversions and stable thermal convection regions over urban areas. In the case of relatively flat topography, this problem is in reasonable condition. In the case of an area like Pittsburgh with many valleys and hills, the data is not available and the fluid mechanics are not fully understood.

10. One must have an understanding of the diffusion of air pollutants in the atmosphere under various meteorological conditions. Many diffusion models for urban areas have been developed and tried. It appears that in regions when the topography is reasonably flat, such models are becoming effective and could be used to decide on minimum cost strategies of implementing air pollution controls.

11. One must have information on technological feasibility of changing various processes to reduce air pollution at the source and/or treatment of exhaust gases from various sources to render the composition of their effluents safe. This is an area that is now getting extensive treatment—both from the point of view of technology and economics. The time constants, however, are long and the interactions with our economy are complex. One must have a rationale method for determining the levels of air pollutants that are tolerable for various time periods for various purposes. We do not yet have enough information on the health effects of low levels of exposure to various air pollutants. Because of the enormous leverage Air Quality Standards have on costs, it is imperative that the level of investment in this activity be greatly increased and that it be made into a crash program.

12. One must have an understanding of how various costs and benefits to the reduction of air pollution are distributed and the implications of these costs and benefits on employment, use of natural resources, international trade, nontariff

trade barriers, balance of payments, etc. We have barely begun to look at questions such as these.

(II) What are the institutional, political, social and financial arrangements that are essential for success? How do these arrangements differ from existing ones?

At the present time, the elements of various political, social, financial, scientific, technical and other institutional arrangements necessary for success are available. That is, we have mechanisms for raising funds, gathering data, passing and enforcing laws, levying taxes, charging fees, etc. What is really lacking is not a Legislative process or an Executive Branch of Government or an awareness of society that air pollution is a problem or a lack of institutions. What is lacking is a mechanism for bringing together a creditably analysed set of policy alternatives for decision makers and proponents for various actions to examine and discuss.

In dealing with the mobile source air pollution we have used, at the Federal level, and in most State Governments, particularly California, and adversary procedure to arrive where we are today. Unfortunately, the process has not led to a least cost, maximum benefit solution, especially since there are no rules of evidence and one of the adversaries is prosecutor, jury and judge. It may be that with mobile source air pollution it was necessary to use whatever mechanisms we had available in order to get started on control of the problem. However, in examining what has taken place in this area, it is clear that we cannot proceed to follow the same methods in the control of all the undesirable effluents associated with operating the modern society.

(III) What impediments in whatever area must be removed before we can succeed?

The science and technology content of the information base required for decision making in mobile source air pollution control is high. It is intimately involved in virtually every aspect of the problem. Many of the adversary groups involved in restoring the quality of our environment take extreme points of view; in some cases, contrary to the Laws of Nature. Unfortunately, the process got off on the wrong foot. Immediately, sides were drawn—the good guys versus the bad guys. This made it extremely difficult for a deliberative evaluation on the pros and cons of various alternative courses of action.

Second, it seemed at least at first that the bad guys were lying. In addition, a number of popular crusaders appeared and appeared to confuse the evidence. Finally, restoring the quality of our environment became a popular political issue causing various groups of policy makers to outdo each other in terms of establishing the appropriate images. None of this led to preparing and examining the kinds of data or doing the analysis necessary to make equitable, cost effective, policy decisions. We do not have, at the present time, an institution creditable to all parties concerned, whose views or analytical results for policy options would be believed by the various parties.

If mobile source air pollution were the only problem to be dealt with, this would not be so serious. It would eventually sort itself out. However, the country as a whole is proceeding to handle, not only its air pollution problems but its land use, solid waste, noise, water pollution, etc. problems in a similar way. The total cost will be outrageous. It could easily lead to an environmental backlash in which the public decides it doesn't want to pay what is required to restore the quality of our environment.

PART FOUR

Social Issues

THE MOTOR VEHICLE AND THE RETURN OF PERSONALIZED TRANSPORT

JOHN G. B. HUTCHINS

Graduate School of Business and Public Administration, Cornell University, Ithaca, N.Y. 14850, U.S.A.

The modern transport revolution has two features—the development of low-cost bulk transport by rail and water and the rise of personalized motor service for packaged freight and passengers. The latter system is now becoming mature in the United States and western Europe. It is setting in motion powerful forces tending to create new equilibria in transportation, business logistics, and personal living patterns. Many new options have become available, especially with respect to land use. The decline of the older pattern based on railroad common carrier service is already far advanced as new routing patterns, costs, and service packages are made available. Many economic functions formerly clustered in the city center can now be more advantageously conducted in rapidly widening metropolitan rings, and even beyond. The age of the non-rail-connected industry, the mobile family, and even the mobile home is upon us. The end result will be a reduction of congestion in the city, new patterns of surburban development, and new qualities of living, despite the many short run problems.

When at the end of this century the history of transportation is written it will be most surely observed that in the seventies we were in the flood tide of a new and major transportation revolution which was rapidly altering the structure, behavior, and performance of the transportation industries, the logistics of business and government, and the location and style of living. From our point of view this revolution has two major features. The first is the development of cheaper bulk and mass transport the world around by means of giant tankers ranging upwards to over 325,000 dwt., bulk carriers of up to 150,000 dwt., and large, fast, containerships by sea, of multiple-barge pushboat operations capable of moving on inland waters as much as 25,000 tons of coal in one movement, and jumbo rail cars of 90 tons or more capacity and unit trains overland. This development includes new and more efficient mechanisms and procedures for programming, controlling, and safeguarding shipments. The second feature is the appearance of personalized transport by truck and automobile on such a scale as to nearly blanket the United States and western Europe. It is this second aspect with which we are concerned.

THE MOTOR VEHICLE AND THE NEW PERSONALIZED TRANSPORT PACKAGES

We begin by considering the transport package which is relevant. Clearly it it much more than basic transportation at a rate per ton-mile or passenger mile. Rather the relevant package for each user is the optimum mixture for him of rate or cost, routing, schedule, reliability, equipment features, and flexibility. The automobile and truck, operating on a modern highway system, have been able to offer new, attractive, and varied packages as compared to those previously offered by rail and steamship common carriers. These latter have been finding much difficulty in meeting this competition in the package freight and passenger markets. In broad terms the new package can be called *personalized transport*. It is primarily a product of a matured motor carrier system.

The motor transport system has reached widely different levels of maturity in various areas, and thus provides different proportions of personalized transport in the total mix. To be regarded as mature a motor system should exhibit the following characteristics:

(1) A system of four-lane, limited-access highways interconnecting all significant nodal points and capable of permitting the unrestricted operation of vehicles at optimum speed and efficiency, backed up by other surfaced rural highways in such density that very few places and areas lack effective access to the national economy.

(2) A metropolitan expressway system connecting the centre with inter-city routes, and connect-

ing various sections with each other by belt-ways and other cross-town routes.

(3) The existence of a motor-carrier enterprise structure capable of meeting service demands at costs and rates consistent with the inherent technical capabilities of motor transport.

(4) A ready supply of various types of equipment suitable for the road system and the traffic requirements.

(5) An adequate, well distributed supply of supporting facilities, including filling stations, fuel suppliers, repair shops, and roadside motels and restaurants.

(6) A system of public agencies and private firms capable of building, financing, and repairing the highway system and of controlling traffic thereon.

In only a few areas is the system reasonably mature. At that stage a high proportion of families is no longer primarily dependant on common carrier rail and bus service, and a large proportion of the mercantile, manufacturing, and service enterprises, except for those having bulk movements in or out or mass production, no longer finds it necessary to locate with reference to railroad or steamship facilities. Today a not uncommon sight is a substantial, low, modern plant located along a major inter-city highway. This freedom for individuals and entrepreneurs is producing important new patterns.

The United States is approaching this stage. The four-lane, limited-access, high-speed Interstate System, which will ultimately total 42,000 route miles, now has some 32,000 miles completed. On these routes trucks carrying over 40,000 lb of freight can travel hour after hour at 60 m.p.h. or better for as long as the driver is allowed to operate each day. A motorist with several drivers on board can conceivably go coast to coast in not much over 48 h. This system is backed up by the older state highway system plus other good surfaced roadbed totalling some 2,800,000 miles.[1] Automobile registrations are approaching 90 million, or 1 car for about 2·3 persons. Inter-city movement of manufactured products, by tonnage, excluding coal and petroleum, is over 50% by truck, although overall in ton-miles trucks have 21%. Vacation travel is perhaps over 83% by automobile. Roadside service is a growth business. There are of course many problems, but the question is what locational patterns will newly mobile businesses and living units adopt in a relatively free environment.

Not all areas even remotely approach this state.

South and Central America have hardly begun their development except for areas near to ports and capital cities.[2] The potential for economic development by modern highway construction here is very great, but vehicles and fuels are often costly. Soviet Russia is predominantly in the railway age.[3] Most elements of developed motor transport are absent from much of Asia and Africa. In some under-developed nations there has been a debate as to whether in the face of limited resources to rebuild and expand an older railroad system or to build a highway system. The results can be predicted to differ, with the former tending to encourage heavy and often extractive industry, exports, and economic nodality at ports and capital cities, and the latter to open hitherto non-market-oriented areas, to encourage light industry, crafts, and agriculture, and to create general local mobility. We shall, however, confine our attention to trends in mature motor systems such as the United States.

As the transport revolution moves forward there is a necessity for all involved to find a new equilibrium. Actually there are two facets to this adjustment. Shippers and travellers are continually engaged in efforts to adapt their business logistics and personal life patterns to the new transport parameters as well as to other pressures. In this there are economic, social, and political aspects. In the short run this adjustment means optimization based on the users' present facilities and patterns. It often involves putting up with much congestion, inefficiency, and discomfort, but the result is often still preferable to a still older rail-oriented adjustment. In the long run—at least several decades—there is a different adjustment, involving new locational patterns for business and government, new residential arrangements, and often new life styles. In such adaptations competitive pressures, capital availabilities, the economics of obsolescence, institutional restrictions, local public policies, and ideals of life styles play a part. The general effect of the motor vehicle is decentralization, but other forces, notably the generalization of electric power and the telephone have been also significant factors pointing the same way. The question is what kind of system, given considerable long-run fluidity, will the motor vehicle tend to create given time.

There is, however, a second type of movement toward a new equilibrium, namely that in transportation itself. This is particularly true in the United States, where some 347 railroads and 3700 major regulated motor carriers struggle for economic life, not to mention some 12,000 small regulated motor carriers. It would be a great mistake to assume that rates and service are not affected by changes in the logistical patterns of shippers. Historically, the railroads have tended to take the lead, but since motor carriers compete they must also respond. Competition among carriers may be parallel, circuitous, and directional. The rivalry of gateways and their respective supporting carriers also is often acute. For example, grain can move, from the midwest overseas via the North Atlantic, Great Lakes, Gulf, and sometimes Pacific Coast parts. If forced, a rail carrier may be expected not to quit until rates fall until it can no longer cover its long-run marginal cost. Each carrier may be the promoter of particular producer interests in major markets. Since railroads often have marginal costs well below average costs there can be considerable scope for adjustments. The result is a complex of groupings, blankets, key points, long-short haul situations, classification adjustments, and special commodity rates. Within the limits imposed by their higher marginal costs truckers are similarly involved. These reaction patterns in turn influence the shipper adjustments.

HISTORICAL PATTERNS OF LONG RUN LOGISTICAL EQUILIBRIUM

To set the problem in perspective let us consider some of the equilibria which have occurred during the course of past transport revolutions. Two patterns are of interest: the adjustment to road, water, and canal transport in Europe and America in the century or more before 1830, and the adjustment to the railroad and steamship age just before World War I. The former was the last age of personalized transport, but one in which costs, risks, transit times, and uncertainties were very high. The latter was a period almost entirely dominated by railroad common carriage and by its phenomena of rate wars, monopolistic pricing, personal, place, and commodity discrimination, strategic maneuvers to secure traffic, empire building, port and gateway rivalries, speculation, and both large profits and bankruptcies.

The overland transport in the former period was by pack horse and wagon, generally over locally maintained roads of quite uncertain quality. The building of turnpikes in Britain and the United States, beginning in the eighteenth century, did quite a bit to improve some routes. So did the improvements in surfacing associated with Mc-Adam and Telford. Costs were very high, being in the 30–40 cent per ton-mile range at the end of the eighteenth century in the United States.[4] By sea, where a great mass of small sailing ships in the 50–300 gross-ton range served the trades, costs were far lower. At this stage inputs of food and drink, which have been estimated at 600–700 lb *per capita*, generally outweighed both other inputs and the outputs of towns, and hence agglomeration was attracted toward areas of high agricultural output such as the plains of Flanders and the west of England.[5] Except by river and sea long distance bulk movements presented very high cost. By land common carrier services were few. Merchants normally handled their own movements both by land and sea. While it is dangerous to generalize, it seems reasonable to conclude that costs were roughly proportional to weight and distance if allowance is made for terrain and obstacles by land, and for winds, currents, and weather conditions by sea. Nevertheless, by roads, paths, and waterways nearly all points in Europe, and most of these in the settled portions of America, were tied in to the regional economies by this system.

The logistics of the economy which this system tended to create can be described in general terms as follows. (1) Because of problems of food supply agglomeration was low, with some exceptions at water transport and political centers. (2) Communities traded manufactured goods for agricultural products in town markets whose radii were limited. Considerable price variations could persist between markets, as is evidence for instance by local famines in France in the sixteenth and seventeenth centuries. (3) Within the "business district" transport was by foot and animals and was almost entirely personalized. This amounted to high but short range mobility which was favorable to the centralizing of community institutions. Longer range trade tended to be limited to the lighter, higher valued products by land, and to these and some foods and materials by sea. The full development of this fascinating adjustment, which created the often still beautiful towns of Europe and early America, is beyond the scope of this paper.

The transport adjustment caused by the railway age almost completely reversed this earlier one, and created a new one based on economies of scale in industry and transport, high nodality with respect to business functions, much higher levels of agglomeration, and transit systems to integrate the enlarged communities. Its monuments have been the central railroad station, the steel skyscraper, concentrated freight yards, dock systems, and rail roadbeds. It replaced personalized transport with common carrier service, and made location with reference thereto of extreme importance. There is grave doubt that this pattern can survive fully intact with the weakening of the common carrier and spread of a new roadway network.

Its primary achievement was the building of the railroad net. Between 1831 and 1870 the United States built 53,000 miles, and Europe 64,000 miles. By then the shape of things was changing rapidly. Then between 1871 and 1914 the former built another 187,000 miles and the latter 143,000 miles. The United States peak mileage of 254,000 was achieved in 1916. Not all of this mileage was high-grade trunk line. It is interesting to compare these figures with the soon-to-be completed United States Interstate motorway system of 42,000 miles. Some of this system is roughly parallel to major rail routes, but important other segments provide in effect new linkages. It is also interesting to note that beyond America and Europe in no other continent did the railroad age become mature in the sense that all significant centers, ports, raw materials, sources, and producing areas were interlocked. The railway age is far from concluded, at least in the United States, but the carriers are now seeking a new equilibrium with motor and water transport involving new patterns of specialization and cooperation.

Let us focus attention on those aspects of the railway logistic equilibrium which are most characteristic of an oligopolistic structure without serious outside challenge and having a mix of low out-of-pocket costs, joint costs, common costs, and economies of traffic density. The history of course should include much beside the analysis of industrial organization in economists' terms, particularly the evolving concepts of the entrepreneurs, local pressures for a "place in the sun", state and federal grants, the influence of land settlement, and the lures of power and speculative profit. North has shown that by 1900 costs per ton mile had dropped from the 30–40 cent level of the wagons a century earlier to around 1 cent per ton mile.[6] Furthermore, the transport package was strikingly improved in respect to speed, reliability, schedules, types of equipment, and service features. Even on the Mississippi the picturesque steamboat traffic passed its peak in the 1860's, not to revive until the 1930's under the influence of waterway improvements, diesel pushboats, and the multiple barge

system. There was little competition from road transport. By 1921 the American highway system totalled some 3,000,000 miles, but less than 500,000 miles were regarded as improved. Much road traffic was either local or a feeder to the railroads. Not until 1914 did the production of motor vehicles exceed that of carriages. This transport system was almost entirely a common carrier one, both in inter- and intracity service. Except for some coastwise schooners, and river craft, personalized transport had vanished in intercity traffic. The new adjustments of business and individuals had to be based on this fact.

Certain features of the railroad rates and service favored concentration both in metropolitan areas, and within these in the city centers. Because of rail and rail-steamship rivalries, ports and nodal points tended to have lower rates and better service than others. The availability of routings to markets was greater. These places therefore become favored locations for trading and warehousing. On the other hand, coal, the great weight-losing raw material, attracted much industry to such coal districts as Pittsburgh and the Ruhr. Both metropolitan and industrial agglomerations could now be cheaply supplied over large distances with foods. As population concentrations increased so did that in retailing, represented especially by down-town department stores. The down-town section, because of its communications and ready contacts, also became an essential location for the administration of the new giant enterprises. Thus was born the characteristic steel skyscraper. Besides executive and staff offices these were inhabited by many business auxiliaries. The focus of personal life was the railroad station, and of business activity the plants, rail yards, and docks. Inside the city a transit system was required to tie it together. Increasingly, city workers found it costly or impossible to maintain the style or ideal of life they desired, much of which was inherited from a small town past, and hence began to move out to suburban locations along the rail lines, thus initiating commuting. The overall effect was greatly to accentuate site value differentials, both rural and urban. This trend has been reversed by modern motor transport.

NEW LEVELS OF MOBILITY AND THEIR SIGNIFICANCE

We now come to consider how motor transport, passenger and freight, has altered the earlier pattern and will continue to do so. Clearly, it has done so in many ways—economic, political, and social. The rate of change has been most rapid wherever the transport and economic structures have been most flexible, and capital and other resources have been readily available for highways, building, and industrial facilities. Particularly vulnerable have been the old down-town business district, in-town retailing, the skyscraper office complex, the old-style seaport, and some patterns of commodity distribution. Also much involved are many rail freight services, the rail commuting system, city transit, the inter-city rail passenger service, and traditional ocean break bulk freight systems. The entire pattern is in process of long range change which will go far.

The mature motor vehicle age means a return to personalized transport at a far higher level of productivity, and a partial but by no means full reversal of the nodal tendencies of the railroad age. The options of shipper and travellers have been greatly widened. The number of geographical locations available to manufacturers and merchants has been drastically increased. Within 60 miles or so of a metropolitan center almost any location on a good highway has become a possible site. The shipper has the choice of common, contract, or private truck service. By piggyback the competitive range of rail carriers has been extended, and likewise motor common carriers find new advantages in rail service. Locations on rail branch lines are of much less value for many, but not all, firms, and such lines are rapidly being abandoned. In effect nearly perfect competition in transport has replaced monopoly or oligopoly wherever rail rates have been above normal motor carrier costs. Private motor service has been the fastest growing segment in the United States, and its costs in effect have placed a ceiling on value-of-service rates. But over and above this fact the motor truck, whether for hire or private, has freed the shipper from the tyranny of the large trainload and infrequent schedule so dear to efficiency-minded railroaders. Indeed, the conflicts between the sales and operating departments of rail carriers has become especially acute as traffic has been lost to trucks under the pressure of operating economies of train size. The end result has been a massive relocalization of economic activity.

In passenger movements the influence of personalized service is most evident. In the United States the automobile population has changed from 1 vehicle for 13 persons in 1920 to 1 for approximately 2 persons at present. Registrations in 1969 totalled 87 million. The average yearly mileage per vehicle is close to 10,000. Except where traffic congestion, bad roads, parking difficulties, and long business trips are involved the automobile is the universally favored mode for business, shopping, personal, and recreational travel. The mobile family is a common feature. Others have access to vehicles through the expanding car rental systems. Only the automobile offers the package of ready availability, unrestricted routing, nominal cost for additional persons, baggage, and packages up to capacity, and, in many cases, superior speed door to door. Only air service can compete with the highway on the longer hauls, and this often involves automobile use at each terminal. Thus we have reached the stage of the mobile businessman and the mobile family, and through the trailer we are seeing the onset of the mobile home. As *per capita* income rises, leisure increases, and economic life becomes more decentralized the automobile presents an unbeatable front to railroad passenger service, railroad commuting, and even short-haul air service, in which terminal delays eat up much of the advantage in speed.

What further changes may be created by the maturing of the motor vehicle age to the point at which individuals, families, and businesses have nearly complete mobility and flexibility, whether by private vehicle of for-hire service? It appears that they will be most evident in: (1) the organizational structure and rate and service patterns of for-hire transport; (2) the nodal position of the central business district; (3) business logistics, including plant locations, inventories, and distribution systems; (4) the pattern of business activity in the widening outer rings of many centers; (5) residential development; (6) the role and function of seaports; and (7) the style and quality of life both at home and at work.

In the United States, which has a private but extensively regulated transportation system, the adjustment process in transport is well advanced. Overall trucks have been handling over half of the loadings of manufactured products excluding petroleum products and coal. From the railroad point of view the villain is the rapidly expanding private trucking. Some corporations operate as many as 5000 units on extensive schedules between their several plants and outward to retailers. The pressure of the motor vehicle is causing railroads to eliminate much branch line mileage, to concentrate on heavy duty service to major producers and to bulk freight shippers, to rationalize by mergers which permit of the elimination of duplicate mileage, terminals, and services, but also reduce the geographical scope of railroading, to compete for private truck, common carrier truck, and ocean container traffic by means of special trailer-or container-on-flat-car service, to speed up operation, and to slowly cut back on rates at the high end of the scale. Middlemen have proliferated. Many railroads are uncertain as to whether or not they should restrict piggyback business to rail-owned trailers at rail rates. In the end the modern railroad system could consist of heavy-duty line-haul service for large loads, and piggyback operations which could haul trailers or containers on major routes in volume at well below driving cost. The railroad would thus be the wholesaler and the truck the retailer. In the United States many railroads would like to become transportation companies, but the existing regulatory system tends to keep the modes separate and competitive. Despite rail mergers the level of competition has clearly increased except at the bulk freight level, but even for this business inland water transport has become more extensive and competitive, partly with the aid of trucks. The burden of adaptation faced by railroad management has thus been more severe than at any previous time, while the adaptation of the regulatory system moves but slowly.

The common carrier motor lines for their part have been engaged in the formation out of a disorganized mass of over 20,000 small carriers some years ago a strongly anchored structure of regional and national carriers, some of which rival major railroads in size and traffic. These carriers operate many routes parallel to major rail lines, but they also provide service on route patterns not existing in the railway age. The route systems of the large common carriers frequently represent the accidents of "grandfather" route awards, acquisitions of other carriers, and grants of new routes. Neither the rail nor the motor carrier industry route pattern appears to have optimum characteristics. The end result, however, is a far wider range of services, geographically speaking, and a wider range of service qualities. There are many opportunities for further adaptation.

A calculation of the Department of Transportation of the distribution of the United States transportation bill of 1970, estimated at 161 billion of 1965 dollars, shows how far the motor vehicle age has progressed. In percentages, taking the total bill at 100, domestic expenditure was 96·99, divided into highway 81·14, rail 6·86, water 2·39, air 5·82, and pipeline 0·78. Of the highway component the figures are passengers 40·67 (divided into automobiles 39·24, and bus 1·43), freight 22·38, and nonfreight truck 18·09. The automobile total breaks down by use into private 31·19, business 5·94, government 1·12, and taxi 0·99, and the bus total into school 0·45, inter-city 0·31, and other 0·08. The truck total is divided nearly equally between local 10·99, and inter-city 11·27. It is interesting that this last figure was divided into for-hire 7·02, and private 4·24. All percentages are based on the national total. The percentages for metropolitan transportation are rail transit 0·20, rail commuter 0·09, and bus local transit 0·60, showing the dominance of the latter over rail.

METROPOLITAN AREAS AND THE NEW BUSINESS AND PERSONAL LOGISTICS

The effect of motor transport on the central business district has been and will be severe. We have to ask what functions formerly conducted there can be better performed where more space is available, rents are lower, buildings are less costly, and personnel can be more readily secured. One such activity is the wholesale trade, which formerly clustered around rail yards and docks, but has now in such cities as Boston mainly moved out to ring and radial highways.[7] Another is much light manufacturing, which can ship by truck even if some incoming supplies are by rail. A third is retailing, which has clearly moved to the metropolitan ring shopping centers to be close to housewives and to secure lower cost plants with ample parking. A fourth is business staff activity—data processing, bank clearing, reservation control, accounting, and research. In some instances American firms are finding the multi-structure campus-design attractive. A fifth is top executive activity, some of which is moving toward ring roads, though much remains still downtown. But automobile transport, the telephone, and other means of communication, and in some cases access to major airports, have made such ring locations possible. Following these changes we also are seeing the decentralization of retail banking and personal service activities. There is left in the central business district primarily corporate executive activity, and that of its corps of supporting specialists, major national banking, money market and organized commodity market activity, and a certain amount of specialized retailing. Increasingly the female shopping population there is declining, with serious effects on urban transport. There is a tendency for in-town workers to move out. Conversely, as the central business district becomes more inhabited by day by administrative and professional types there is a tendency for high grade living units to go up; thus accentuating the visible gap between the well paid and the immobile groups in remaining slum areas. Because of bad load factors urban transport has become a financial headache. With increased leisure the desire for access to recreational areas becomes for many a further decentralizing force. How far these trends will go depends on many rigidities, but they are well advanced. At some point a new equilibrium in the inner city will emerge.

In terms of business logistics the motor truck has made it feasible to supply many customers from large, mechanized warehouses, often located in the country along major highways. Rail service is usually desirable for incoming material. A characteristic pattern in many industries consists of separate specialized production facilities for components, assembly plants, and well located distribution centers. The single story building suitable for mechanized operation is favored. Workers may drive over 50 miles on fast roads to reach their employment. This system is replacing highly concentrated operations involving dense agglomeration. Of necessity speed of change is slow.

The decentralization of these aspects of economic life is of necessity also decentralizing all types of economic activity which are either consumer oriented or enterprise oriented. Hence there is a still further outward drift of repair shops, medical services, amusements, professional personnel, educational establishments, and even governmental agencies.

With respect to seaports the prime requirements for containership services is ample space for sorting boxes, and ready access to major highways and trunk line rail service. Many older type facilities are unsuitable. Thus much of the general cargo business of New York is moving from the older Manhattan and Brooklyn docks across the Hudson to the new terminals at Newark, Elizabeth, and Staten Island. Estimates are that within a decade some 80% of this trade will be containerized. Thus the older waterfronts may be expected to become

esplanades or areas for high rise apartments. At the same time the economics of ship operation and the availability of motor feeder service are concentrating ship services in a few ports.

Finally and perhaps the most striking change in the United States is the spread of residential developments, which were formerly close in, out across agricultural lands as fas as 70 or 80 miles from the central business district. The ideal of many a family is clearly the single house set on its own lot. The motor vehicle has increased drastically the speed of commuting, while at the same time increasing the suitable home-site areas. Furthermore, the abovementioned decentralizing trends permit of still wider living arcs for those working in the metropolitan ring. At its extreme there are such planned living centers as Reston and Columbia in the Washington outer rings.

This revolution is obviously bringing better living conditions for many in terms of space, privacy, facilities, mobility, convenience, and cost. On the other hand it is putting severe pressure on many who have to live in decaying older cities. There are problems of pollution and congestion. Hopefully the former can be solved by technology and government action. The latter will be aided by the mere process of decentralization. For the great majority, however, the motor vehicle has widened the horizon of life and opened up new opportunities, while at the same time creating new hazards.

REFERENCES

1. For a good statement of the position of motor transport in the United States, with various statistics see John B. Rae, *The Road and Car in American Life*. M.I.T. Press, Cambridge, Mass. (1971).
2. Willian H. Dodge, "Network analysis of central American regional highway system" *Transportation Research Forum Papers*, 1969. George W. Wilson, *The Impact of Highway Investment on Development*. Brookings Institution, Washington (1966).
3. Ernest W. Williams, *Freight Transportation in the Soviet Union*. National Bureau of Economic Research, Princeton (1962).
4. Douglas C. North, "The role of transportation in the economic development of North America" In *Les Grandes Voies Maritimes dans le Monde*, p. 222. S.E.V.P.E.N., Paris (1965).
5. W. H. Dean, *Theory of the Geographic Location of Economic Activities*. Ann Arbor, Mich. (1938).
6. North, *op. cit. (supra)*.
7. See Rae, *op. cit. (supra)*, Chap. 12 for a particularly good analysis of developments in American metropolitan areas.

RECREATION AND CARS

GEOFFREY WALL

Department of Geography, University of Sheffield, Sheffield, S10 2TN, U.K.

This paper considers the role played by the car in influencing both pleasure trip and holiday patterns, and its consequent effect upon the structure of the holiday trades. It is suggested that while the car-owner may contribute to congestion and environmental damage, the car also provides a means by which visitors to vulnerable areas may be managed. The importance of positive planning for outdoor recreation is stressed and several examples of planned provision for recreation are outlined.

INTRODUCTION

Numerous studies indicate the important recreational role of the car.[1-5] Long summer traffic jams on major roads leading to the coast and to the more accessible rural beauty spots bear witness to this fact. The Pilot National Recreation Survey,[6] Sillitoe's survey of leisure,[7] and Masser's case study of South Birmingham[8] indicated that car-owning households report a higher level of participation than those without cars over almost the whole spectrum of outdoor recreation activities. Lickorish has summarized the situation as follows:[4]

If you have a car you are more likely to take a holiday; more likely to spend more time away from home; more likely to become a mobile man or mobile woman, rather than a home spender. You will spend a greater proportion of your income outside your home and outside the complex trade servicing the home. You will spend it in the travel trades.

Comparatively high rates of participation among car-owners, the noise and fumes emitted by the car, its parking requirements, and its intrusion in the landscape mean that the car-owners tend to have a greater impact on the environment than those without cars. The number of cars has been rising more rapidly than population growth with the result that not only are there more cars on the road, but an increasing proportion of families are owning cars, and a rising proportion of households are acquiring a second vehicle (Table I). It must be recognized that there will always be families who do not own cars, and there will always be car-owners who, on occasion, prefer to leave their vehicles in the garage. The recreational requirements of such people should not be neglected. However, all trends indicate that the recreation pattern of the future will be dominated by car-owners, even if their chosen activities do not always involve the use of a car.

TABLE I
Private cars in Great Britain, 1949–1969

Year	1949	1954	1959	1964	1969
Number of cars (millions)	2·1	3·1	5·0	8·2	11·2
Population (millions)	48·4	49·4	50·5	52·5	54·0
% of households with no car	86	81	76	63	49
% of households with one car	14	19	23	33	44
% of households with two or more cars	0	0	1	4	7

Source: Road Research Laboratory.

THE CAR AS A RECREATIONAL OPPORTUNITY

Almost all car-owners derive pleasure from driving and they are very reluctant to use alternative means of transport to the car: only 8·7% of a sample of Hull (U.K.) car-owners claimed not to enjoy driving and only 3·9% did not use a car on their last pleasure trip.[10] The car may be viewed as a catalyst of growth in outdoor recreation in that it has diminished the friction of distance between home and the recreation site and has thereby been one factor encouraging participation in outdoor activities of almost all types. One of the few exceptions to this generalization is cycling which has tended to decline in popularity in Britain as road congestion has increased, although in New York where provision has been made for cyclists in Central Park, this activity has expanded in recent months.

However, the car is much more than simply a transport medium and it may be viewed as a recreation opportunity in its own right. The journey in the car often contributes as much to the enjoyment of the outing as events at the trip destination and recent surveys suggest that driving for pleasure is the most popular recreation activity in both Britain[11] and North America.[12] The car also tends to act as a focus of activity once a destination has been reached and the majority of trippers can satisfy their recreational needs within a few yards of a parked car or by remaining inside it. For example, Wager found that 26% of visitors to common land remained within their car and a further 20% sat or picnicked near their car.[13] Similarly, 17% of visitors to Box Hill, Surrey, did not leave their car and a further 50% picnicked beside the car. Burton found that the car was the focal point of activity for 75% of all groups visiting Box Hill.[14] In the case of Ashdown Forest, Furmidge found that 83% of picnicking activities occurred immediately adjacent to motor vehicles and 16% within 6 yd.[15] These figures suggest that many people are seeking a visual rather than a physical contact with the countryside. In consequence the greatest pressures on the environment are concentrated where there is access for cars and there is evidence of the erosive forces of feet and wheels acting together where large numbers congregate in the same small area.[16] The impact of mass recreation declines rapidly with distances from routeways and parking places and at only a short distance from the road their effects may be minimal. Thus there are large areas of the countryside where recreation pressures are small. It has been shown that visitors to both Snowdonia[17] and Dartmoor National Parks[18] tend to congregate in a limited number of recreation foci leaving much of the area of the parks relatively free from visitor pressure. In many such circumstances the concentration of visitors and the soil and vegetation changes which they induce may be less damaging and more manageable than less intensive usage over a wider area. Much here will depend upon the nature of the resource.

HOLIDAYS

The increased use of the automobile is influencing the character of holidays. Since 1951 the propor-

tion of all holidaymakers travelling to their main holiday destinations in Britain by rail has fallen from almost one half to less than one fifth, and in the same period the relative importance of bus and coach transport has also declined markedly. These modes of passenger transport have suffered severely in competition with the car which now carries two thirds of all holidaymakers to their main holiday destinations (Table II).

TABLE II
Method of transport to main holiday destinations in Great Britain, 1951–1968 (%)

Year	1951	1955	1962	1966	1968
Car	27	34	54	64	66
Bus/coach	27	33	18	20	16
Train	47	37	26	16	14
Other	0	0	0	7	5

Source: British Travel Association, *Patterns in British Holiday-making 1951–1968*. London (1969).

The expansion of car travel has induced a growth in touring holidays which is having repercussions upon the structure of the holiday trades. The holiday industry in Britain has been traditionally geared to stays measured in units of 1 week and the majority of holidays begin and end on a Saturday. The increase in the popularity of touring holidays is forcing the industry to adapt to meet the rising demand for shorter overnight stops. These tend to be less profitable because of increased costs of service and administration, and demand is less reliable since tourists are less likely to book accommodation in advance.

There has also been a gradual change in the types of accommodation utilized because the car-owner often takes his accommodation and food with him. This is reflected in the growing popularity of self-catering holidays. Much of this increase may be attributed to the rising number of caravans. In 1951 caravans were insufficiently significant in the total holiday picture to merit consideration as a separate accommodation category by the British Travel Association, but by 1968, 16% of all main holidays taken by British persons in Britain were in caravans.[19] It should be noted that the majority of caravans are "static", but, nevertheless, considerable sanitary problems are created by the indiscriminate use of lay-bys by a comparatively small number of mobile caravanners, especially on major roads approaching holiday areas.

Mobility from campsite to campsite coupled with the ease of transporting equipment by car probably accounts for the comparatively high proportion of campers among those with cars. Camping, at 10·2% of last main holidays, was found to be more than twice as popular among a sample of Hull car-owners than among the holidaymakers of the nation as a whole.[20]

Since most car-owning holidaymakers utilize their cars to reach their holiday destinations in Britain and are mobile once their holiday location has been reached, their influence is felt in a wider area than a single resort. The most popular form of holiday is to stay at one resort but to take trips to places of interest in the surrounding area: 46·6% of Hull car-owners opted for an arrangement of this sort on their last main holiday. Only 28·4% of the sample spent their entire holiday in one resort and a quarter spent nights in more than one location, i.e. took a touring holiday.[21] In consequence it is realistic to see the distribution of holidays in the context of holiday regions rather than within the confines of a large number of individual resorts.

PATTERNS OF MOVEMENT

So far this paper has emphasized changes brought about in outdoor recreation by mass car-ownership but there is a strong conservative element even in the recreation habits of car-owners. In spite of the potential mobility which the motor car affords the majority of car-owners spend most of their leisure relatively near their homes. The characteristic of weekend and evening recreation is that time is relatively limited. This being the case, the value of the car lies not so much in the distance which it allows its occupants to travel, but in its flexibility of timing and use. For instance, it has been shown that over half of a sample of Hull car-owners last pleasure trips were to destinations within 30 miles of Hull and three-quarters were less than 50 miles distant.[22] Similarly, day tripper traffic to the North-East coast is highly localized. For instance, 90% of the day visitors to the Durham coast live within the county, and most of the movement is within a 20-mile catchment area. At Whitley Bay 70% of visitors come from towns less than 16 miles away, and over 50% are from the adjacent Tyneside conurbation.[23] It appears that the car has enabled the recreational belt of cities to be used more intensively but has not greatly extended its radius.

More adventurous journeys in terms of distance might be expected in connection with holidays but even holiday demand has an important local component. Despite the impression of the Lake District as a national holiday area, visitor hinterlands of Keswick, Ambleside and Windermere exhibit a marked regional emphasis. In each of the three resorts almost half the staying visitors come from northern England, the main source areas being south-east Lancashire, the North-East and the

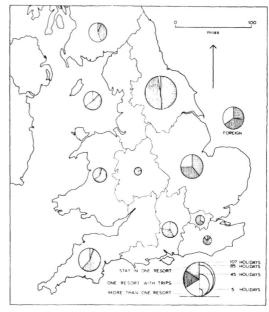

FIGURE 1 The regional distribution of main holidays taken by a sample of 500 Hull car-owners.

West Riding of Yorkshire.[24] More than a quarter (26·5%) of a sample of Hull car-owners spent their last main holiday in the local North-East holiday region as defined by the British Travel Association (Figure 1), and the proportion of their secondary holidays in that region was even greater (40·7%).[25] Touring holidays are popular with a substantial minority of Hull car-owners but it would be erroneous to regard car-owners as uniformly rushing from place to place on their holidays. The majority still stay overnight at only one resort. The movements of most holidaymakers are still relatively localized and the major pattern of holiday travel is to places within the car-owners region of residence or into adjoining regions.

IMPLICATIONS FOR PLANNING

The automobile gives the car-owner a greater choice of both holiday and pleasure trip destinations than if his movements were restricted by the availability of public transport. The public transport network, particularly with the contraction of rural bus and train services, has a predominantly urban orientation and has always served the seaside resorts better than the rural areas. This is one reason why seaside resorts have attracted a large proportion of pleasure seekers. However, the automobile frees its occupants from the confines of the well-served routes and enables car-owners to push more deeply and in greater numbers into previously inaccessible rural areas. Such are the pressures that this can create that the question of cars and car-parking in attractive rural areas is regarded as "the biggest single planning problem connected with countryside leisure".[26] Improvement of the antiquated roads in rural areas would only encourage more visitors and would, at the same time, remove one of the attractions which draw visitors from the towns. Such are the pressures of cars upon the countryside that it is being increasingly suggested that cars should be completely excluded from some vulnerable areas of high amenity value by the adoption of traffic-free zones.[27] This, by itself, without alternative provision for the car-borne recreationist elsewhere, will not solve the problem but will merely shift its location for the car-owner will find somewhere where he and his family can play.

It is paradoxical that while traffic-free zones are designed to alleviate the problems of recreation reception areas many of the local residents are often opposed to the implementation of such measures. It is feared that freedom of movements of residents might be curtailed by the difficulty of distinguishing between local and "foreign" traffic and traffic-free zones might discourage the tourist and thereby restrict income from that source. Many attractive rural areas are marginal in terms of agriculture and catering for the tourist by the provision of farm holidays, overnight accommodation, and caravan and camping sites could provide a supplementary source of income to residents of some problem rural areas, particularly those near the major tourist routes.[28,29]

A point that is usually overlooked is that although the car-borne recreationist often contributes to congestion and environmental damage, the car also provides a means by which visitors to vulnerable areas may be managed. Since most pleasure seekers are car-borne, and since most of them are loath to move very far from their cars, the regulation of traffic flows and parking affords an opportunity to adjust the magnitude and location of the incidence of outdoor recreation. Thus, for instance, at times of peak demand a one-way traffic system is imposed in Farndale in the North York Moors National Park. Increasing attention is being given to the use of car parks and parking fees as a management tool in rural planning.[30,31] One example of a policy of this type has been adopted as an experiment by the Peak Park Planning Board in the Goyt Valley, Derbyshire (U.K.) (Figure 2). A motorless zone

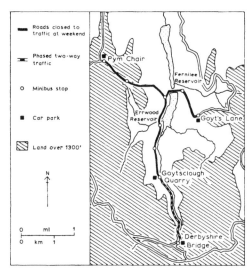

FIGURE 2   The Goyt Valley Experiment, Derbyshire.

covers some 5 miles of road leading to, and around, the Errwood Reservoir, which is a major point of attraction for motorists in the summer. New car parks are the start of a mini-bus service which operates throughout the motorless zone, and the cost of which is included in the parking fee. Public reaction to this scheme appears to be favourable.

It should be emphasized that it would be wrong to adopt a negative approach to planning for outdoor recreation. Planning for recreation in Britain has lagged behind North America where a more positive approach has often been adopted and recreation facilities have been created to cater for car-borne demand. For example, the "metroparks" of the Huron–Clinton Metropolitan Authority, serving the greater Detroit area of the U.S.A., were created purely to provide recreation outlets for the residents of that city.[32] When one park is full motorists are directed along a network of freeways to other parks which are less congested. Other examples of North American recreation planning are scenic driveways such as the Blue Ridge Parkway, which is over 500 miles in length, and the Gatineau Parkway near Ottawa in Canada. These parkways are designed specifically for pleasure travel, with driving speeds controlled and parking places and viewpoints provided. There are few examples of such drives in Britain but a preliminary venture of the Forestry Commission indicates that they could also be popular on this side of the Atlantic.[33] North American examples of recreation provision may not be capable of direct translation to British conditions but they do serve as examples of what can be achieved by positive planning.

The major form of planned recreation provision in Britain evolves around the provision of Country Parks.[34] A Country Park is "an area of land, or land and water, normally not less than 25 acres in extent, designed to offer to the public with or without charge, opportunity for recreational activities in the countryside".[35] Grants of up to 75% of approved expenditure are available to both individuals and local authorities for the construction of such facilities. The aims of the policy are to make it easier for those seeking recreation to enjoy their leisure in the open, without travelling too far and adding to congestion on the roads; to ease pressure on the more remote and solitary places; and to reduce the risk of damage to the countryside. An example of what is being attempted may be seen at Elvaston Castle, Derbyshire where the castle is being restored and the surrounding park landscape is being modified by new planting and the creation of picnic glades. Toilets, a cafe, and vehicular access including a car and coach park to accommodate five hundred cars and twelve coaches are being constructed at a cost of £43,250.[36] At the time of writing Country Park policy is still in its infancy and it is too early to evaluate the degree to which its aims are being realized.

CONCLUSIONS

The car has enabled the present generation to do more things and see more places than earlier generations found possible. Pressure on the countryside and coastline for recreational use by an increasingly numerous, leisured and mobile population is steadily intensifying.[37,38] Associated with an increasing demand for recreation outlets is also a greater diversity in the types of demand. In other words, along with a quantitative change a qualitative change is also taking place. Without effective planning and management more trespass, traffic congestion and environmental damage is likely to result. It is essential that outdoor recreation is recognized as a land use in its own right. Facilities must be designed and provided specifically for outdoor recreation if increasing and varied demands are to be met and if, at the same time, conflicts with other uses of land and water are to be kept to a minimum. Without such positive planning and provision there is a real danger that the motorist may destroy that which he has come to seek.

ACKNOWLEDGEMENT

The author is extremely grateful to Dr. Patrick Lavery who made many useful comments upon an initial draft of this paper.

REFERENCES

1. T. L. Burton, *Windsor Great Park: A Recreation Study, Studies in Rural Land Use Report No. 8.* Wye College, Ashford (1967).
2. T. L. Burton and G. B. Wibberley, *Outdoor Recreation in the British Countryside, Studies in Rural Land Use, Report No. 5.* Wye College, Ashford (1967).
3. B. Cracknell, "Accessibility to the countryside as a factor in planning for leisure" *Regional Studies* **1**, 147–161 (1967).
4. J. R. Duffell and G. R. Goodall, "Worcestershire and Staffordshire recreational survey 1966" *J. Town Planning Institute* **55**, 16–23 (1969).
5. W. B. Yapp, *The Weekend Motorist in the Lake District.* H.M.S.O., London (1969).
6. British Travel Association and University of Keele. *Pilot National Recreation Survey* (2 Vols). London (1967 and 1969).
7. K. K. Sillitoe, *Planning for Leisure.* Government Social Survey, H.M.S.O., London (1969).
8. I. Masser, "The use of outdoor recreation facilities" *Town Planning Review* **37**, 41–54 (1966).
9. L. J. Lickorish in British Travel Association, *Conference on the Travel and Holiday Industry in Yorkshire.* Harrogate (1963).
10. G. Wall, Patterns of Recreation of Hull Car-owners. Unpublished Ph.D. Thesis, University of Hull (1970).
11. B. Cracknell, *op. cit.*
12. Outdoor Recreation Resources Review Commission, *Study Reports.* U.S. Government Printing Office, Washington D.C. (1962).
13. J. F. Wager, "Outdoor recreation on common land" *J. Town Planning Institute* **53**, 398–403 (1967).
14. T. L. Burton, "A day in the country: a survey of leisure activity at Box Hill in Surrey" *J. Royal Institute of Chartered Surveyors* **98**, 378–380 (1966).

15. J. Furmidge, "Planning for recreation in the country-side" *J. Town Planning Institute* **55**, 62–67 (1969).

16. C. Buchanan and Partners, *South Hampshire Study.* H.M.S.O., London (1966).

17. J. W. E. H. Gittins, "Recreation pressure in the Snow-donia National Park". Paper presented at the Annual Meeting, British Association for the Advancement of Science, Swansea (1971).

18. Joint School Survey, "People at play in Dartmoor National Park" *Geographical Magazine* **42**, 266–279 (1970).

19. British Travel Association, *Patterns in British Holiday-making 1951–1968.* London (1969).

20. G. Wall, *op. cit.*

21. *Ibid*

22. G. Wall, "Seasonal variations in pleasure trip patterns" *Recreation News Supplement* **5**, 19–23 (1971).

23. P. Lavery, *Patterns of Holidaymaking in the Northern Region, Research Series No. 9.* Department of Geography, University of Newcastle-upon-Tyne (1971).

24. *Ibid.*

25. G. Wall, *op. cit.* (1970).

26. E. Beazley, *Designed for Recreation.* Faber, London (1969).

27. D. Rubinstein and C. Speakman, *Leisure, Transport and the Countryside, Fabian Research Series No. 277.* Fabian Society, London (1969).

28. T. L. Burton, *Outdoor Recreation Enterprises in Problem Rural Areas, Studies in Rural Land Use Report No. 9.* Wye College, Ashford (1967).

29. E. T. Davies, *Tourism and the Cornish Farmer.* Depart-ment of Economics, Exeter University (1969).

30. Countryside Commission, *Methods of Charging at Rural Car Parks.* London (1969).

31. G. Runnicles, "Lepe car parking experiment" *Recreation News Supplement* **4**, 22–25 (1971).

32. J. A. Patmore, *Land and Leisure.* David & Charles, Newton Abbot (1970).

33. Countryside Commission, *Scenic Drive Survey Dovey and Gwydyr Forests July 1969.* London (1970).

34. J. A. Zetter, *The Evolution of Country Park Policy.* Countryside Commission, London (1971).

35. *Policy on Country Parks and Picnic Sites.* Countryside Commission, London (1969).

36. A. Latham, "Focus on Elvaston country park" *Recrea-tion News Supplement* **1**, 8–10, 24 (1970).

37. R. J. S. Hookway, *Planning for Leisure and Recreation.* Countryside Commission, London (1968).

38. M. Dower, "Leisure—its impact on man and the land" *Geography* **55**, 253–260 (1970).

PROBLEMS OF THE POOR IN AN AUTO OWNING ENVIRONMENT

ROBERT E. PAASWELL†

*Engineering and Applied Sciences, State University of
New York at Buffalo, New York, U.S.A.*

The growth of urban areas coupled with the high rate of increase in car ownership have caused a true segregation between those with access to a car and those without. Especially hard hit are the poor, who find that the alternatives available to them are few, expensive or unsuitable for their trip needs. In the latter category much transportation that is within physical reach is truly ineffective as time and cost penalties are excessive. A short term solution for Buffalo, New York, in the form of a dial-a-bus system is briefly described.

The pattern of urbanization in the United States is a phenomenon that can be attributed in great part to the influences of auto ownership and the corresponding commitment to a national highway program necessary to support the use of the car. The 1970 census shows that as movement from rural to urban areas continues unabated, at the expense of rural areas, movement within the urban areas is also occurring. This movement is from center city to suburb, with the result that in 1970 more people lived in the suburban areas‡ of U.S. cities than in the central areas. This familiar movement has its origin in the availability of housing, at reasonable prices, in suburban areas coupled with rapid access to employment and shops located throughout the metropolitan areas (no longer confined to the Central Business District (CBD)) and, as mentioned, high rates of car ownership. In 1970 more than 79% of American households had one or more cars. In fact, nearly 90% of the personal trips by all modes made in the United States were made by members of these car-owning households, because they had cars.

Because of this commitment to the car, public transit has been allowed to decline until finally in the late 1960's and early 1970's, horrified planners recognized that poor policy planning decisions had to be undone. This should not be seen however as a change in the planners commitment to the car. The current rush to provide mass transportation systems is predicated on the facts that rush hour congestion by cars must be relieved, and that high volumes of demand in peak corridors must be served in more efficient ways. The design of the solutions to these problems, where, theoretically four out of every five people had access to a car, was oriented in favour of the car owners. It is a commonplace by now that this orientation to the car cannot solve all transportation problems. It is those problems, not addressed by planning for the automobile, which this paper will discuss, both in reference to the U.S. and the U.K.§

CAR-ACCESSIBILITY

There are two key factors that limit car use. The first is the ability to drive (as usually shown by licensing requirements), and the second is to have a car at your disposal when you wish to travel. It is a gross oversimplification to conclude that four out of five people have access to a car at any given time simply because four out of five households have cars. First in most households, not everyone has a license, or chooses to drive. This is obviously true for those under 16, and to a great extent true for the elderly. Secondly, when a car is in use by one member of a household, it is not therefore necessarily in use by all members of the household. This is shown by the fact that average car occupancy is less than two persons per trip, while average household size is nearly four persons.

---

† Currently on sabbatical leave at Political and Economic Planning, London, U.K.

‡ Fifty-five per cent of SMSA's lived in suburban areas as opposed to the 45% who lived in central cities.

§ Although many of these very problems have in fact caused the past and current emphasis on the car, we don't deal directly with those causes here.

In addition, one household in five is without a car—amounting to nearly 45 million people or almost the population size of Great Britain. Clearly, the automobile is not the universal solution to travel needs that so many people claim it to be. Of this latter group, and even for the non-drivers in car owning households the hardest hit by problems of inadequate transportation are the poor. Unfortunately for planners, and more unfortunately for the poor, they do not constitute a homogeneous group with similar and easily identifiable demands and needs.

The most uniform characteristic of the poor is the low car ownership. In both the U.K. and U.S. there is a predictable direct relationship between household income and car-ownership. As household incomes decline, the probability of owning a car declines. In the U.S. at $3000/year only one family in two have a car. This decreases to one in four at $1000/year. In the U.K. one in two families owns a car at income levels of £1500 ($3750) and one in four families own a car at £1000 ($2500). At the equivalent of £1000/year in the U.K. ($2500), only one family in 20 own a car. However, at £2500 four of every five families own at least one car. Car ownership is not uniformly distributed. In dense areas of Greater London, for example, car ownership levels are approximately 15% less than in low density areas of similar income. In the U.S. in the largest central cities, 57% of the households run at least one car. In suburban and adjacent areas more than 85% of the households are car owning.

Ownership varies with age as well. In the U.S., of families where heads are 65 or over, 56% own one or more cars. This is due to both lower incomes at higher ages, and decreasing ability to drive. Income and expenditure surveys in both the U.S. and the U.K. indicate that the elderly make up a large segment of the urban poor. However, because each group has its special trip needs it is essential to distinguish between the elderly and other groups of urban poor as each group.

THE POOR AND TRAVEL NEEDS

The urban poor can be subdivided by age, occupation, or a variety of other social characteristics. For the purposes of examining travel needs, the poor can generally be grouped into households with employed head and those with unemployed head; and then to further divide these groups into elderly and younger members. The fundamental means for doing this are to:
(1) establish the need for work trips;
(2) establish the need for essential but non-work trips; and food, medical etc.;
(3) establish the need for all other discretionary trips.

Surveys have shown that nearly twice as many trips are made daily by car owning households than by non-car owning households. And, of course as car ownership is linked to incomes poorer families make fewer vehicle trips than richer ones.

For working households, a certain percentage of the income must be allocated for the work trip. As the job is necessary to sustain income, a percentage of the income must be budgeted for work trips, to ensure accessibility to work. For example, at $3000/year, a head of household might actually take home $50/week. Unable to afford a car or operate it in most urban areas, the principal worker would either walk to work or take public transit. While walking to work would be the cheapest solution, rapid decentralization of industry, together with the growth of white collar service industries in the CBD's where the poorer generally live, mean that the walk trip is not always possible. Less than 4% of the total work trips in the U.S. are walk trips, and the bulk of these are made by clerical, labouring or service workers. Public transit in the U.S. outside a very few of the largest areas is subsidised at very low levels (if at all). Hence fares are high. With typical one-way bus fare in the U.S. in an ungraded zone of $0.40, a worker would be spending (at $50/weekly income) 8% of his income on work travel. When he has alloted the remainder of his income for food and rent little remains for any other trips. Therefore, at this income level, trips for medical and shopping purposes become discretionary trips, as opposed to the more common definition of discretionary trips as leisure trips and leisure-shop trips.

Other daily needs that must be satisfied include food-shopping, shopping for other necessary goods, some personal business (banking), medical trips and a limited number of social visits. Low income households are frequently located in high density urban areas, where such services are likely to be found in greater number. In high density areas it is possible to walk for the most essential services (food, drugs, schools, banks, post office), with no expenditure on travel.

In London, it is still not uncommon to walk to work. In some of the inner areas more than 25%

of the work trips are walk trips. The availability of a wide variety of goods in High Streets shopping areas and street markets make it possible to have most services within 20 min walk in most inner areas. In the U.S., growth of the supermarkets has closed many neighbourhood grocers and the trend in urban areas is toward store clusters that are more accessible by car than by walking for most people. In addition, the car has had an impact on the walk trip in another way—pedestrian safety. Disregard for pedestrians by vehicles, coupled with more segregated areas, or lack of sufficient marked crossing areas, make it unpleasant and physically difficult to walk. In fact it has become part of planning policy, to consider the pedestrian as an obstacle to the car, as a limit to adequate flow, rather than vice versa. Again, the poor who have the fewest cars lose the most. The walk trip is most essential because of income limitations as noted above.

The 1962 London Traffic Survey showed not surprisingly that lower income groups made fewer vehicle trips per household than did higher income groups. In non-car owning households at income levels of £1000, three trips per household were a daily average, whereas at £2000, 5·71 trips per household were made. While the figures do not include walk trip, it must be concluded that this difference in trip rates recorded by the survey can be attributed to: more walk trips at lower income levels; and/or more non-work trips at higher income levels; or simply trips desired but not made. A fundamental problem on the latter category is that the poor often forego trips because transportation is unaccessible to them.

THE NATURE OF TRANSPORTATION ALTERNATIVES

The variety of transportation alternatives available to the poor depend particularly upon their lifestyle, and this is in turn affect by three major factors: income level, age and (particularly in U.S. urban areas) race. The need for transport is clearly a function of the daily activities of the individuals in a household; which activities are finally carried out is a function of the priority level placed upon them, and that is quite often a function of income. The 1970 Family Expenditure Survey for the U.K. established spending patterns as a function of income level. In Table I, transport allocations are clearly seen to be proportioned to income.

TABLE I
Income-transport Relationships

Nominal	Per cent expended	Per cent of transport expenditures on car
less than 10	6·0	60
10–15	6·7	55
15–20	9·0	70
20–25	9–8	75
25–30	11–0	76

Source: Great Britain Family Expenditure Survey/1970, H.M.S.O. 1971.

Further expenditure breakdown illustrates that approximately 50–55% of the income at all income levels are allocated to housing, food and transport. However, as income decreases, the percentage spent on housing increases. The lower percentage spent on transport at lower income levels therefore must occur in part because of the increased percentage referred for housing. This can be seen clearly in figures given in the same survey for the elderly, a significant group of poor, who had an average weekly income (1970) of £18·23, allocating 17% to housing, 27% to food and only 8% to transport. While car ownership among the poor is at a much higher level in the U.S. than the U.K. it is still at levels low enough to make car trips unavailable to more than half the population. Because of the costs of operation, the use of car pools for journeys to work with shared expenses is quite common for U.S. In many urban areas, where blue collar or service industries have become suburbanized or dispersed, the car is often the only way to reach the job. In the case of car pools this leads to a serious problem of access for riders when the driver is unable to make the trip.

INEFFECTIVE TRANSIT

It is generally presumed that physical presence of some type of transit system within an acceptable distance from potential riders is evidence of accessibility. However, this is not always the case, for a variety of factors may conspire to make even a physically present, nearby system either actually inaccessible or of no use to the rider. Time presents real penalties when the trip time exceeds the limits of what the traveller considers worthwhile. The time of the trip can be budgeted as a percent of the total daily activities, or as a percent of the total daily travel time for all purposes. For example, a survey, in Buffalo, N.Y., of low-income workers indicated that only 39% would be willing to travel more than 60 min by public transit to work. The average journey to work in Buffalo is 18 min by car, 30 min by public transit.[†]

In an urban environment, where the most likely mode of travel is by car, there is also some social stigma to riding on a bus.

The point has been made above that low income (or lack of enough income to cover all necessary trips other than work trips) can prevent the use of a transport system no matter now near it is. There is another inhibiting factor, directly related to, and a refinement of, the income factor. As work,

$$\text{Penalty as a percentage of local costs} = \frac{(\text{Costs locally} - \text{costs elsewhere}) - \text{trip cost}}{\text{Costs locally}} \, 100\%$$

shopping areas, and medical areas become more dispersed, it is less likely that one bus trip will satisfy a variety of trip purposes. In most U.S. cities the flat fare is standard, reduced fares, passes and tokens are uncommon. Thus, it is necessary for the rider to have the necessary money for each trip he wishes to make at the time the trip is made. This is in contrast to car operation where fixed amounts are paid on a more or less periodic basis and are not usually allocated for each trip.

Finally, public transit, although readily accessible by walk from the home may just not go where it is needed. In many U.S. urban areas, the lower income groups are clustered near the core of the city. Public transit routes, especially those geared to satisfying peak travel demands, have concentrated on C.B.D.—suburban routes. As the job market for unskilled labour has left the core area and become more scattered in suburban areas, the journey to work has become difficult without a car. In a recent survey[1] in Buffalo, N.Y., taken in an inner-city district, it was noted that of all work trips made to the suburban areas, 85% were by car. This was in an area in which car ownership

was less than one for every two households. The survey again underlined the important role of the car pool, as bus service was perceived to be inadequate. In an area of high unemployment, (twice the national average), residents of the survey area felt that time and trip cost penalties of inadequate public transport were too great to overcome.[‡]

ADDITIONAL COST PENALTIES

Ineffective transportation creates additional cost penalties for those affected by it. These are the cost penalties that are paid to obtain necessary services that are not available in a reasonable travel time. They can be assessed on a trip purpose basis and are most severe on food shopping. A penalty formula can be given as:

It can readily be seen that as long as the difference in costs are greater than the trip cost their is some penalty to pay. For example, if shopping costs in an inner city neighbourhood store are 10% higher than in a suburban market, for a $20 (£8) grocery bill, and an 80c (32p) return fare, there is a 6% penalty that can be assessed for inadequate transit if the suburban trip cannot be made. Again, the survey in Buffalo showed that 40% of car owners shopped out of their neighbourhood, while only 20% of non-car owning households did their shopping out of the neighbourhoood. Poor transportation just reinforces certain disadvantages that are borne by low-income households, the lack of access to adequate opportunities for most activities being a clear example. The fact that non-car owning households made shopping trips less frequently than car owning households with similar incomes further underlines the inadequacies of available transit.

A SOLUTION

Planning over the next few decades must make a commitment to the solution of the mobility problems of the poor. The problems of the poor

[†] Trip time travelled by public transit is also a measure of available opportunities. In Buffalo, for work trips, in the 18 min by car 200,000 jobs would be available. In the same bus time only 50,000 jobs are available, or to encounter the same number of opportunities by bus 45 min of travel are necessary.

[‡] Statistics of the study supported those perceptions by showing trip time and costs relative to income to be significantly greater than the norm in the metropolitan area.

must be attacked immediately and not be protracted by the uncertainties and indecisions of long range planning. Only in this way can we stop the current momentum of urban physical change caused by increasing car ownership. The alternative is to create further polarization between car owning and non-car owning groups.

One short range solution is currently being studied in Buffalo, New York in an experiment to provide free rides to a group of elderly residents. The Buffalo Model Cities Agency, supported by the City of Buffalo and the Federal Government through funds of the Model Cities Program is supporting a program of planned services, including transportation, in an inner area of Buffalo that is characterized by low income, a high percentage of substandard housing, low car ownership (one-half of the metropolitan area average), a low index of job skills and high unemployment. Utilizing limited funds, the Agency established a dial-a-bus experimental service aimed at the elderly (over 59) residents of the area. This group was chosen because the community felt that of all those in the area handicapped by poor transportation, the elderly were the most severely hit. Not only were their incomes lower than the median for the area thus reducing their travel budget, but conditions of climate, such as long severe winters, prevented their making even simple pedestrian trips.

While the target population was theoretically 7000 elderly residents, reductions due to advanced age, infirmity, or no desire to travel at all made the true target population approximately 5000–5500. Four, 10-passenger Dodge van-buses were equipped with two-way radios and air conditioning, and special steps and painted with identifying marks so they would readily be identified by the community. The service was provided free for any trip purpose. To utilize the service, which operates between 8 a.m. and 11.30 p.m. (23.30) the requirements were that a rider had to be a resident of the area and over 59. To get a ride, the resident called in a day in advance, giving his name, address, phone number, the time he wanted to be picked up and his destination. He also states whether he is handicapped, needs assistance, and if anyone is accompanying him.

A dispatcher takes the call, books it on a scheduling sheet made out for each bus and attempts to fill the ride at the desired time. If the time cannot be met the rider is given suggested alternative times, usually within several minutes of the requested time.

Assessment of the data after several months of service (the service was initiated in January 1971 and is still ongoing) has led to an accurate description of the travel patterns of the elderly. Twenty-eight per cent of the trips were for medical purposes, twenty for shopping and the remainder divided over a variety of activities. This illustrates the relative importance of medical visits in the life of the elderly. Travel was concentrated in two main periods of the day, 9–12 and late afternoon. Although service was available, after 6 p.m. (18.00) less than 10% of the trips were taken at that time. The number of riders on the four vehicles is currently over 200 passengers per day and slowly increasing.

While this service was provided as a short term solution, immediate acceptance by the community has shown that new transportation alternatives are needed, and the distribution of trips shows that they are needed to satisfy necessary activities. Greater levels of mobility will then permit more nondiscretionary trips to be made, thus reducing the gap between car owning and non car-owning households.

REFERENCE

1. R. Paaswell et al., Mobility of Inter-city Residents. Dept. of Civil Engineering, State University of New York, Buffalo (Dec. 1970).

AUTHOR AND TITLE INDEX